A Handbook of
Statistical
Analyses
using Stata

Second Edition

D0552687

A Handbook of
Statistical Analyses
using Stata

Second Edition

Sophia Rabe-Hesketh
Brian Everitt

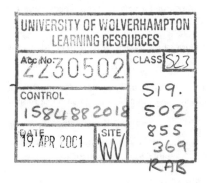
CHAPMAN & HALL/CRC

Boca Raton   London   New York   Washington, D.C.

Library of Congress Cataloging-in-Publication Data

Rabe-Hesketh, S.
  A handbook of statistical analyses using Stata / Sophia Rabe-Hesketh,
Brian Everitt.—2nd ed.
        p. cm.
  Rev. ed. of: Handbook of statistical analysis using Stata. c1999.
  Includes bibliographical references and index.
  ISBN 1-58488-201-8 (alk. paper)
  1. Stata. 2. Mathematical statistics—Data processing. I. Everitt, Brian.
  II. Rabe-Hesketh, S. Handbook of statistical analysis using Stata. III. Title.
QA276.4 .R33 2000
519,5′0285′5369—dc21   00-027322
                                                                             CIP

© 2000 by Chapman & Hall/CRC

No claim to original U.S. Government works
International Standard Book Number 1-58488-201-8
Library of Congress Card Number 00-027322
Printed in the United States of America 1 2 3 4 5 6 7 8 9 0
Printed on acid-free paper

Preface

Stata is an exciting statistical package which can be used for many standard and non-standard methods of data analysis. Stata is particularly useful for modeling complex data from longitudinal studies or surveys and is therefore ideal for analyzing results from clinical trials or epidemiological studies. The extensive graphic facilities of the software are also valuable to the modern data-analyst. In addition, Stata provides a powerful programming language that enables 'tailor-made' analyses to be applied relatively simply. As a result, many Stata users are developing (and making available to other users) new programs reflecting recent developments in statistics which are frequently incorporated into the Stata package.

This handbook follows the format of its two predecessors, *A Handbook of Statistical Analysis using S-Plus* and *A Handbook of Statistical Analysis using SAS*. Each chapter deals with the analysis appropriate for a particular set of data. A brief account of the statistical background is included in each chapter but the primary focus is on how to use Stata and how to interpret results. Our hope is that this approach will provide a useful complement to the excellent but very extensive Stata manuals.

We would like to acknowledge the usefulness of the Stata Netcourses in the preparation of this book. We would also like to thank Ken Higbee and Mario Cleves at Stata Corporation for helping us to update the first edition of the book for Stata version 6 and Nick Cox for pointing out errors in the first edition. This book was typeset using LaTeX.

All the datasets can be accessed on the Internet at the following Web sites:

- http://www.iop.kcl.ac.uk/IoP/Departments/BioComp/stataBook.stm
- http://www.stata.com/bookstore/statanalyses.html

S. Rabe-Hesketh
B. S. Everitt
London

To my parents, Birgit and Georg Rabe
Sophia Rabe-Hesketh

To my wife, Mary Elizabeth
Brian S. Everitt

Contents

Distributors for Stata

The distributor for Stata in the United Kingdom is:

Timberlake Consultants
Unit B3, Broomsleigh Business Park
Worsley Bridge Road
London SE26 5BN

email: info@timberlake.co.uk
Web site: http://www.timberlake.co.uk
Telephone: 44(0)-20-8697-3377

In the United States, the distributor is:

Stata Corporation
702 University Drive East
College Station, TX 77840

email: stata@stata.com
Web site: http://www.stata.com
Telephone: 900-782-8272

For a list of distributors in other countries, see the Stata Web page.

CHAPTER 1

A Brief Introduction to Stata

1.1 Getting help and information

Stata is a general purpose statistics package which was developed and is maintained by Stata Corporation. There are several forms of Stata, "Intercooled Stata," its shorter version "Small Stata," and a simpler to use (point and click) student package "StataQuest." There are versions of each of these packages for Windows (3.1/3.11, 95, 98, and NT), Unix platforms, and the Macintosh. In this book, we will describe Intercooled Stata for Windows although most features are shared by the other versions of Stata.

The Stata package is described in seven manuals (*Stata: Getting Started, Stata User's Guide, Stata Reference Manuals 1-4*, and the *Stata Graphics Manual*) and in Hamilton (1998). The reference manuals provide extremely detailed information on each command while the User's Guide describes Stata more generally. Features which are specific to the operating system are described in the appropriate *Getting Started* manual, e.g., *Getting Started with Stata for Windows*.

Each Stata-command has associated with it a help-file that may be viewed within a Stata session using the help facility. If the required command-name for a particular problem is not known, a list of possible command-names for that problem may be obtained using *search*. Both the help-files and manuals refer to the reference manuals by "[R] command name," to the User's Guide by "[U] chapter number and name," and the graphics manual by "[G] name of entry."

The Stata Web page (http://www.stata.com) contains information on the Stata mailing list and Internet courses and has links to files containing extensions and updates of the package (see Section 1.10) as well as a "frequently asked questions" (FAQ) list and further information on Stata.

The Stata mailing list, *Statalist*, simultaneously functions as a technical support service with Stata staff frequently offering very helpful responses to questions. The Statalist messages are archived at:
http://www.hsph.harvard.edu/statalist

Internet courses, called *netcourses*, take place via a temporary mailing list for course organizers and "attenders"; Each week, the course organizers send out lecture notes and exercises which the attenders can discuss with each other

until the organizers send out the answers to the exercises and to the questions
raised by attenders.

1.2 Running Stata

When Stata is started, a screen opens as shown in Figure 1.1 containing four
windows labeled:

- Stata Command
- Stata Results
- Review
- Variables

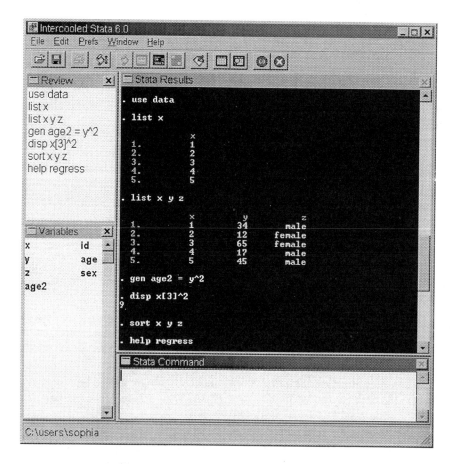

Figure 1.1 *Stata windows.*

A command may be typed in the Stata Command window and executed by pressing the *Return* (or *Enter*) key. The command then appears next to a full stop in the Stata Results window, followed by the output. If the output is longer than the Stata Results window, --more-- appears at the bottom of the screen. Typing any key scrolls the output forward one screen. The scroll-bar may be used to move up and down previously displayed output. However, only a certain amount of past output is retained in this window. It is therefore a good idea to open a log-file at the beginning of a stata-session. Press the button ▨, type a filename into the dialog box, and choose **Open**. If the file-name already exists, another dialog opens to allow you to decide whether to overwrite the file with new output or whether to append new output to the existing file. The log-file may be viewed during the Stata-session and is automatically saved when it is closed. A log-file may also be opend and closed using commands:

```
log using filename, replace
log close
```

Stata is ready to accept new commands when the prompt (a period) appears at the bottom of the screen. If Stata is not ready to receive new commands because it is still running or has not yet displayed all the current output, it may be interrupted by holding down *Ctrl* and pressing the *Pause/Break* key or by pressing the red **Break** button ▨ .

A previous command can be accessed using the *PgUp* and *PgDn* keys or by selecting it from the Review window where all commands from the current Stata session are listed. The command may then be edited if required before pressing *Return* to execute the command. In practice, it is useful to build up a file containing the commands necessary to carry out a particular data-analysis. This may be done using Stata's **Do-file Editor**. The editor may be opened by clicking ▨. Commands that work when used interactively in the command window can then be copied into the editor. The do-file may be saved and all the commands contained in it may be executed by clicking ▨ in the do-file editor or using the command

```
do dofile
```

A single dataset may be loaded into Stata. As in other statistical packages, this dataset is generally a matrix where the columns represent variables (with names and labels) and the rows represent observations. When a dataset is open, the variable names and variable labels appear in the Variables window. The dataset may be viewed as a spread-sheet by opening the **Data Browser** with the ▨ button and edited by clicking ▨ to open the **Data Editor**. See Section 1.3 for more information on datasets.

Most Stata commands refer to a list of variables, the basic syntax being

```
command varlist
```

For example, if the dataset contains variables **x y** and **z**, then

```
list x y
```

lists the values of **x** and **y**. Other components may be added to the command, for example adding **if exp** after **varlist** causes the command to process only those observations satisfying the logical expression **exp**. Options are separated from the main command by a comma. The complete command structure and its components are described in Section 1.4.

Help may be obtained by clicking on **Help** which brings up the dialog box shown in Figure 1.2. To obtain help on a Stata command, assuming the com-

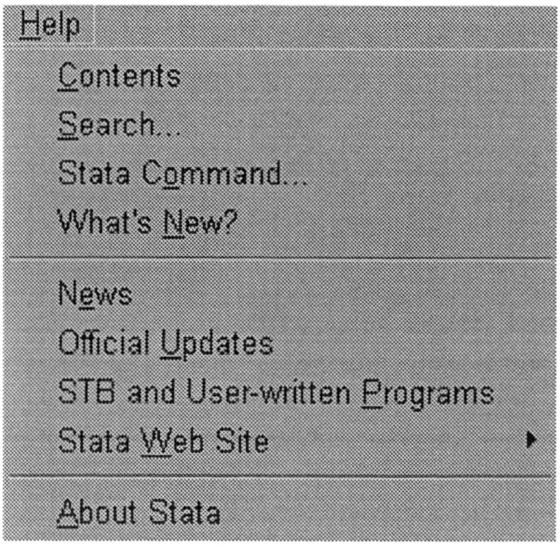

Figure 1.2 *Dialog box for help.*

mand name is known, select **Stata Command...**. To find the appropriate Stata command first, select **Search...**. For example, to find out how to fit a Cox regression, we can select **Search...**, type "survival" and press **OK**. This gives a list of relevant command names or topics for which help-files are available. Each entry in this list includes a green keyword (a hyperlink) that may be selected to view the appropriate help-file. Each help-file contains hyperlinks to other relevant help-files. The search and help-files may also be accessed using the commands

```
search survival
help cox
```

except that the files now appear in the Stata Results window where no hyperlinks are available.

The selections **News**, **Official Updates**, **STB**, **and User-written Programs** and **Stata Web Site** all enable access to relevant information on the Web providing the computer is connected to the internet (see Section 1.10 on keeping Stata up-to-date).

Each of the Stata windows may be resized and moved around in the usual way. The fonts in a window may be changed by clicking on the menu button ▪ on the top left of that window's menu bar. All these settings are automatically saved when Stata is exited.

Stata may be exited in three ways:

- click on the **Close** button ⊠ at the top right-hand corner of the Stata screen
- select the **File** menu from the menu bar and select **Exit**
- type `exit, clear` in the Stata Commands window and press *Return*.

1.3 Datasets in Stata

1.3.1 Data input and output

Stata has its own data format with default extension *.dta*. Reading and saving a Stata file are straightforward. If the filename is *bank.dta*, the commands are

```
use bank
save bank
```

If the data are not stored in the current directory, then the complete path must be specified, as in the command

```
use c:\user\data\bank
```

However, the least error prone way of keeping all the files for a particular project in one directory is to change to that directory and save and read all files without their pathname:

```
cd c:\user\data
use bank
save bank
```

When reading a file into Stata, all data already in memory need to be cleared, either by running `clear` before the `use` command or by using the option `clear` as follows:

```
use bank, clear
```

If we wish to save data under an existing filename, this results in an error message unless we use the option `replace` as follows:

```
save bank, replace
```

If the data are not available in Stata format, they may be converted to Stata format using another package (e.g., Stat/Transfer) or saved as an ASCII file (although the latter option means losing all the labels). When saving data as ASCII, missing values should be replaced by some numerical code.

There are three commands available for reading different types of ASCII data: `insheet` is for files containing one observation (on all variables) per line with variables separated by tabs or commas, where the first line may contain the variable names; `infile` with varlist (free format) allows line breaks to occur anywhere and variables to be separated by spaces as well as commas or tabs; and `infile` with a dictionary (fixed format) is the most flexible command. Data may be saved as ASCII using `outfile` or `outsheet`.

Only one dataset may be loaded at any given time but a dataset may be merged with the currently loaded dataset using the command `merge` or `append` to add observations or variables.

1.3.2 Variables

There are essentially two types of variables in Stata: string and numeric. Each variable can be one of a number of storage types that require different numbers of bytes. The storage types are byte, int, long, float, and double for numeric variables and str1 to str80 for string variables of different lengths. Besides the storage type, variables have associated with them a name, a label, and a format. The name of a variable y can be changed to x using

```
rename y x
```

The variable label can be defined using

```
label variable x "cost in pounds"
```

and the format of a numeric variable can be set to "general numeric" with two decimal places using

```
format x %7.2g
```

Numeric variables

Missing values in numeric variables are represented by dots only and are interpreted as very large numbers (which can lead to mistakes). Missing value codes may be converted to missing values using the command `mvdecode`. For example,

```
mvdecode x, mv(-99)
```

replaces all values of variable x equal to −99 by dots and

```
mvencode x, mv(-99)
```

changes the missing values back to −99.

Numeric variables can be used to represent categorical or continuous variables including dates. For categorical variables it is not always easy to remember which numerical code represents which category. Value labels can therefore be defined as follows:

```
label define s 1 married 2 divorced 3 widowed 4 single
label values marital s
```

The categories can also be recoded, for example

```
recode marital 2/3=2 4=3
```

merges categories 2 and 3 into category 2 and changes category 4 to 3.

Dates are defined as the number of days since 1/1/1960 and can be displayed using the date format %d. For example, listing the variable time in %7.0g format gives

```
list time
```

```
             time
 1.        14976
 2.          200
```

which is not as easy to interpret as

```
format time %d
list time
```

```
              time
 1.  01jan2001
 2.  19jul1960
```

String variables

String variables are typically used for categorical variables or in some cases for dates (e.g., if the file was saved as an ASCII file from SPSS). In Stata, it is generally advisable to represent both categorical variables and dates by numeric variables, and conversion from string to numeric in both cases is straightforward. A categorical string variable can be converted to a numeric variable using the command **encode** which replaces each unique string by an integer and uses that string as the label for the corresponding integer value. The command **decode** converts the labeled numeric variable back to a string variable.

A string variable representing dates can be converted to numeric using the function **date(string1, string2)** where **string1** is a string representing a date and `string2` is a permutation of `"dmy"` to specify the order of the day, month, and year in `string1`. For example, the commands

```
display date("30/1/1930","dmy")
```

and

```
display date("january 1, 1930", "mdy")
```

both return the negative value -10957 because the date is 10957 days before 1/1/1960.

1.4 Stata commands

Typing `help language` gives the following generic command structure for most Stata commands:

```
[by varlist:] command [varlist] [=exp] [if exp] [in range] [weight]
                                             [using filename]  [, options]
```

The help-file contains links to information on each of the components, and we will briefly describe them here:

[by varlist:] instructs Stata to repeat the command for each combination of values in the list of variables `varlist`.

[command] is the name of the command and can be abbreviated; for example, the command `display` can be abbreviated as `dis`.

[varlist] is the list of variables to which the command applies.

[=exp] is an expression.

[if exp] restricts the command to that subset of the observations that satisfies the logical expression `exp`.

[in range] restricts the command to those observations whose indices lie in a particular `range`.

[weight] allows weights to be associated with observations (see Section 1.6).

[using filename] specifies the filename to be used.

[options] are specific to the command and can often be abbreviated.

For any given command, some of these components may not be available, for example `list` does not allow [using filename]. The help-files for specific commands specify which components are available, using the same notation as above, with square brackets enclosing components that are optional. For example, `help log` gives

```
log using filename [, noproc append replace ]
```

implying that [by varlist:] is not allowed and that `using filename` is required whereas the three options `noproc`, `append`, or `replace` are optional.

The syntax for `varlist`, `exp`, and `range` is described in the next three subsections, followed by information on how to loop through sets of variables or observations.

1.4.1 Varlist

The simplest form of `varlist` is a list of variable names separated by spaces. Variable names may also be abbreviated as long as this is unambiguous, i.e., `x1` may be referred to by `x` only if there is no other variable starting on x such as `x` itself or `x2`. A set of adjacent variables such as `m1`, `m2`, and `x` may be referred to as `m1-x`. All variables starting on the same set of letters can be represented by that set of letters followed by a wild card `*`, so that `m*` may stand for `m1 m6 mother`. The set of all variables is referred to by `_all`. Examples of a `varlist` are

```
x y
x1-x16
a1-a3 my* sex age
```

1.4.2 Expressions

There are logical, algebraic, and string expressions in Stata. Logical expressions evaluate to 1 (true) or 0 (false) and use the operators `<` and `<=` for "less than" and "less than or equal to", respectively. Similarly, `>` and `>=` are used for "greater than" and "greater than or equal to". The symbols `==` and `~=` stand for "equal to" and "not equal to", and the characters `~`, `&` and `|` represent "not", "and", and " or", respectively, so that

```
if (y~=2&z>x)|x==1
```

means "if y is not equal to two and z is greater than x or if x equals one". In fact, expressions involving variables are evaluated for each observation so that the expression really means

$$(y_i \neq 2 \& z_i > x_i)|x_i == 1$$

where i is the observation index.

Algebraic expressions use the usual operators `+ - * /` and `^` for addition, subtraction, multiplication, division, and powers, respectively. Stata also has many mathematical functions such as `sqrt()`, `exp()`, `log()`, etc. and statistical functions such as `chiprob()` and `normprob()` for cumulative distribution functions and `invnorm()`, etc. for inverse cumulative distribution functions. Pseudo-random numbers may be generated using `uniform()`. Examples of algebraic expressions are

```
y + x
(y + x)^3 + a/b
invnorm(uniform())+2
```

where `invnorm(uniform())` returns a (different) sample from the standard normal distribution for each observation.

Finally, string expressions mainly use special string functions such as substr(str,n1,n2) to extract a substring starting at n1 for a length of n2. The logical operators == and ~= are also allowed with string variables and the operator + concatenates two strings. For example, the combined logical and string expression

```
("moon"+substr("sunlight",4,5))=="moonlight"
```

returns the value 1 for "true".

For a list of all functions, use help functions.

1.4.3 Observation indices and ranges

Each observation has an index associated with it. For example, the value of the third observation on a particular variable x may be referred to as x[3]. The macro _n takes on the value of the running index and _N is equal to the number of observations. We can therefore refer to the previous observation of a variable as x[_n-1].

An indexed variable is only allowed on the right-hand side of an assignment. If we wish to replace x[3] by 2, we can do this using the syntax

```
replace x=2 if _n==3
```

We can refer to a range of observations using either if with a logical expression involving _n or, more easily, by using in range, where range is a range of indices specified using the syntax f/l (for "first to last") where f and/or l may be replaced by numerical values if required, so that 5/12 means "fifth to twelfth" and f/10 means "first to tenth", etc. Negative numbers are used to count from the end, for example

```
list x in -10/l
```

lists the last 10 observations.

1.4.4 Looping through variables or observations

Explicitly looping through observations is often not necessary because expressions involving variables are automatically evaluated for each observation. It may however be required to repeat a command for subsets of observations and this is what by varlist: is for. Before using by varlist:, however, the data must be sorted using

```
sort varlist
```

where varlist includes the variables to be used for by varlist:. Note that if varlist contains more than one variable, ties in the earlier variables are sorted according to the next variable. For example,

```
sort school class
by school class: summ test
```

give the summary statistics of **test** for each class. If **class** is labeled from 1 to n_i for the ith school, then not using **school** in the above commands would result in the observations for all classes labeled 1 to be grouped together.

A very useful feature of **by varlist:** is that it causes the observation index _n to count from 1 within each of the groups defined by the unique combinations of the values of varlist. The macro _N represents the number of observations in each group. For example,

```
sort group age
by group: list age if _n==_N
```

lists **age** for the last observation in each group where the last observation in this case is the observation with the highest age within its group.

We can also loop through a set variables or observations using **for**. For example,

```
for var v*: list X
```

loops through the list of all variables starting on **v** and applies the command **list** to each member **X** of the variable list. Numeric lists may also be used. The command

```
for num 1 3 5: list vX
```

lists **v1**, **v3**, and **v5**. Numeric lists may be abbreviated by "first(increment)last", giving the syntax 1(2)5 for the list 1 3 5 (not an abbreviation in this case!). The **for** command can be made to loop through several lists (of the same length) simultaneously where the "current" members of the different lists are referred to by **X**, **Y**, **Z**, **A**, **B**, etc. For example, if there are variables **v1**, **v2**, **v3**, **v4**, and **v5** in the dataset,

```
for var v1-v5 \num 1/5: replace X=0 in Y
```

replaces the ith value of the variable **vi** by 0, i.e., it sets vi[i] to 0. Here, we have used the syntax "first/last" which is used when the increment is 1. See **help for** for more information, for example on how to construct nested loops.

Another method for looping is the **while** command which is described in Section 1.9 on programming but may also be used interactively.

1.5 Data management

1.5.1 Generating variables

New variables may be generated using the commands **generate** or **egen**. The command **generate** simply equates a new variable to an expression which is evaluated for each observation. For example,

```
generate x=1
```

creates a new variable called **x** and sets it equal to one. When **generate** is used together with **if exp** or **in range**, the remaining observations are set to missing. For example,

```
generate percent = 100*(old - new)/old if old>0
```

generates the variable **percent** and set it equal to the percentage decrease from **old** to **new** where **old** is positive and equal to missing otherwise. The function **replace** works in the same way as **generate** except that it allows an existing variable to be changed. For example,

```
replace percent = 0 if old<=0
```

changes the missing values in **percent** to zeros. The two commands above could be replaced by the single command

```
generate percent=cond(old>0, 100*(old-new)/old, 0)
```

where **cond()** evaluates to the second argument if the first argument is true and to the third argument otherwise.

The function **egen** provides an extension to **generate**. One advantage of **egen** is that some of its functions accept a variable list as an argument, whereas the functions for **generate** can only take simple expressions as arguments. for example, we can form the average of 100 variables **m1** to **m100** using

```
egen average=rmean(m1-m100)
```

where missing values are ignored. Other functions for **egen** operate on groups of observations. For example, if we have the income (variable **income**) for members within families (variable **family**), we may want to compute the total income of each member's family using

```
egen faminc = sum(income), by(family)
```

An existing variable can be replaced using egen functions only by first dropping it using

```
drop x
```

Another way of dropping variables is using **keep varlist** where **varlist** is the list of all variables not to be dropped.

1.5.2 Changing the shape of the data

It is frequently necessary to change the shape of data, the most common application being grouped data, in particular repeated measures. If we have measurement occasions j for subjects i, this may be viewed as a multivariate dataset in which each occasion j is represented by a variable xj and the subject identifier is in the variable subj. However, for some statistical analyses we may need one single, long, response vector containing the responses for all occasions for all subjects, as well as two variables subj and occ to represent the indices i and j, respectively. The two "data shapes" are called wide and long, respectively. We can convert from the wide shape with variables xj and subj given by

```
list
```

	x1	x2	subj
1.	2	3	1
2.	4	5	2

to the long shape with variables x, occ, and subj using the syntax

```
reshape long x, i(subj) j(occ)
list
```

	subj	occ	x
1.	1	1	2
2.	1	2	3
3.	2	1	4
4.	2	2	5

and back again using

```
reshape wide x, i(subj) j(occ)
```

For data in the long shape, it may be required to collapse the data so that each group is represented by a single summary measure. For example, for the data above, each subject's responses can be summarized using the mean, meanx, and standard deviation, sdx, and the number of nonmissing responses, num. This can be achieved using

```
collapse (mean) meanx=x (sd) sdx=x (count) num=x, by(subj)
list
```

	subj	meanx	sdx	num
1.	1	2.5	.7071068	2
2.	2	4.5	.7071068	2

Since it is not possible to convert back to the original format, the data may be preserved before running collapse and restored again later using the commands preserve and restore.

Other ways of changing the shape of data include dropping observations using

```
drop in 1/10
```

to drop the first 10 observations or

```
sort group weight
by group: keep if _n==1
```

to drop all but the lightest member of each group. Sometimes it may be necessary to transpose the data, converting variables to observations and vice versa. This may be done and undone using **xpose**.

If each observation represents a number of units (as after collapse), it may sometimes be required to replicate each observation by the number of units, num, that it represents. This may be done using

```
expand num
```

If there are two datasets, *subj.dta*, containing subject specific variables, and *occ.dta*, containing occasion specific variables for the same subjects, then if both files contain the same sorted subject identifier **subj_id** and *subj.dta* is currently loaded, the files may be merged as follows:

```
merge subj_id using occ
```

resulting in the variables from *subj.dta* being expanded as in the **expand** command above and the variables from *occ.dta* being added.

1.6 Estimation

All estimation commands in Stata, for example **regress**, **logistic**, **poisson**, and **glm**, follow the same syntax and share many of the same options.

The estimation commands also produce essentially the same output and save the same kind of information. The stored information may be processed using the same set of *post-estimation commands*.

The basic command structure is

```
[xi:] command depvar [model] [weights], options
```

which may be combined with **by varlist:**, **if exp**, and **in range**. The response variable is specified by **depvar** and the explanatory variables by **model**. The latter is usually just a list of explanatory variables. If categorical explanatory variables and interactions are required, using **xi:** at the beginning of the command enables special notation for **model** to be used. For example,

```
xi: regress resp i.x
```

creates dummy variables for each value of x except the first and includes these dummy variables as regressors in the model.

```
xi: regress resp i.x*y z
```

fits a regression model with the main effects of x, y, and z and the interaction x×y where x is treated as categorical and y and z as continuous (see help xi for further details).

The syntax for the [weights] option is

```
weighttype=varname
```

where weighttype depends on the purpose of weighting the data. If the data are in the form of a table where each observation represents a group containing a total of freq observations, using [fweight=freq] is equivalent to running the same estimation command on the expanded dataset where each observation has been replicated freq times. If the observations have different standard deviations, for example, because they represent averages of different numbers of observations, then aweights is used with weights proportional to the reciprocals of the standard deviations. Finally, pweights is used for inverse probability weighting in surveys where the weights are equal to the inverse probability that each observation was sampled. (Another type of weights, iweight, is available for some estimation commands mainly for use by programmers.)

All the results of an estimation command are stored and can be processed using post-estimation commands. For example, predict may be used to compute predicted values or different types of residuals for the observations in the present dataset and the commands test, testparm, and lrtest can be used to carry out various tests on the regression coefficients.

The saved results can also be accessed directly using the appropriate names. For example, the regression coefficients are stored in global macros called _b[varname]. In order to display the regression coefficient of x, simply type

```
display _b[x]
```

To access the entire parameter vector, use e(b). Many other results may be accessed using the e(name) syntax. See the "Saved Results" section of the entry for the estimation command in the *Stata Reference Manuals* to find out under what names particular results are stored.

1.7 Graphics

There is a command, graph, which may be used to plot a large number of different graphs. The graphs appear in a Stata Graph window which is created when the first graph is plotted. The basic syntax is **graph varlist, options** where **options** are used to specify the type of graph. For example,

```
graph x, box
```

gives a boxplot of x and

```
graph y x, twoway
```

gives a scatter-plot with y on the y-axis and x on the x-axis. (The option twoway is not needed here because it is the default.) More than the minimum number of variables may be given. For example,

```
graph x y, box
```

gives two boxplots within one set of axes and

```
graph y z x, twoway
```

gives a scatterplot of y and z against x with different symbols for y and z. The option by(group) may be used to plot graphs separately for each group. With the option box, this results in several boxplots within one set of axes; with the option twoway, this results in several scatterplots in the same graphics window and using the same axis ranges.

If the variables have labels, then these are used as titles or axis labels as appropriate. The graph command can be extended to specify axis-labeling, or to specify which symbols should be used to represent the points and how (or whether) the points are to be connected, etc. For example, in the command

```
graph y z x, s(io) c(l.) xlabel ylabel t1("scatter plot")
```

the symbol() option s(io) causes the points in y to be invisible (i) and those in z to be represented by small circles (o). The connect() option c(l.), causes the points in y to be connected by straight lines and those in z to be unconnected. Finally, the xlabel amd ylabel options cause the x- and y-axes to be labeled using round values (without these options, only the minimum and maximum values are labeled) and the title option t1("scatter plot") causes a main title to be added at the top of the graph (b1(), l1(), r1() would produce main titles on bottom, left and right and t2(), b2(), l2(), r2() would produce secondary titles on each of the four sides).

The entire graph must be produced in a single command. This means, for example, that if different symbols are to be used for different groups on a scattergraph, then each group must be represented by a separate variable having nonmissing values only for observations belonging to that group. For example, the commands

```
gen y1=y if group==1
gen y2=y if group==2
graph y1 y2 x, s(dp)
```

produce a scatter-plot where y is represented by diamonds (d) in group 1 and by plus (p) in group 2. See **help graph** for a list of all plotting symbols, etc. Note that there is a very convenient alternative method of generating y1 and y2. Simply use the command

```
separate y, by(group)
```

1.8 Stata as a calculator

Stata can be used as a simple calculator using the command **display** followed by an expression, e.g.,

```
display sqrt(5*(11-3^2))
```

$$\boxed{3.1622777}$$

There are also a number of statistical functions that can be used without reference to any variables. These commands end in **i**, where i stands for *immediate command*. For example, we can calculate the sample size required for an independent samples t-test to achieve 80% power to detect a significance difference at the 1% level of significance (2-sided) if the means differ by one standard deviation using **sampsi** as follows:

```
sampsi 1 2, sd(1) power(.8) alpha(0.01)
```

```
Estimated sample size for two-sample comparison of means

Test Ho: m1 = m2, where m1 is the mean in population 1
                  and m2 is the mean in population 2
Assumptions:

         alpha =    0.0100  (two-sided)
         power =    0.8000
            m1 =         1
            m2 =         2
           sd1 =         1
           sd2 =         1
         n2/n1 =      1.00

Estimated required sample sizes:

            n1 =        24
            n2 =        24
```

Similarly, **ttesti** can be used to carry out a t-test if the means, standard deviations and sample sizes are given.

Results can be saved in local macros using the syntax

```
local a=exp
```

The result may then be used again by enclosing the macro name in single quotes ` ´ (using two separate keys on the keyboard). For example,

```
local a=5
display sqrt(`a´)
```

$$\boxed{2.236068}$$

Matrices can also be defined and matrix algebra carried out interactively. The following matrix commands define a matrix a, display it, and give its trace and its eigenvalues:

```
matrix a=(1,2\2,4)
matrix list a
```

```
symmetric a[2,2]
        c1   c2
r1    1
r2    2    4
```

```
dis trace(a)
```

$$\boxed{5}$$

```
matrix symeigen x v = a
matrix list v
```

```
v[1,2]
        e1   e2
r1    5    0
```

1.9 Brief introduction to programming

So far we have described commands as if they would be run interactively. However, in practice, it is always useful to be able to repeat the entire analysis using a single command. This is important, for example, when a data entry error is detected after most of the analysis has already been carried out. In Stata, a set of commands stored as a do-file, called for example, *analysis.do*, can be executed using the command

```
do analysis
```

We strongly recommend that readers create do-files for any work in Stata, i.e., for the exercises of this book.

One way of generating a do-file is to carry out the analysis interactively and save the commands, for example, by selecting **Save Review Contents** from the menu of the **Review** window. Stata's **Do-file Editor** can also be used to create a do-file. One way of trying out commands interactively and building up a do-file is to run commands in the Commands window and copy them into the **Do-file Editor** after checking that they work. Another possibility is to type commands into the **Do-file Editor** and try them out individually by highlighting the command and clicking into **Tools - Do Selection**. Alternatively, any text-editor or word-processor may be used to create a do-file. The following is a useful template for a do-file:

```
/* comment describing what the file does */
version 6.0
capture log close
log using filename, replace
set more off

command 1
command 2
etc.

log close
exit
```

We will explain each line in turn.

1. The "brackets" /* and */ cause Stata to ignore everything between them. Another way of *commenting out lines* of text is to start the lines with a simple *.

2. The command **version 6.0** causes Stata to interpret all commands as if we were running Stata version 6.0 even if, in the future, we have actually installed a later version in which some of these commands do not work anymore.

3. The **capture** prefix causes the do-file to continue running even if the command results in an error. The **capture log close** command therefore closes the current log-file if one is open or returns an error message. (Another useful prefix is **quietly** which suppresses all output, except error messages).

4. The command **log using filename, replace** opens a log-file, replacing any log-file of the same name if it already exists.

5. The command **set more off** causes all the output to scroll past automatically instead of waiting for the user to scroll through it manually. This is useful if the user intends to look at the log-file for the output.

6. After the analysis is complete, the log-file is closed using `log close`.

7. The last statement, `exit`, is not necessary at the end of a do-file but may be used to make Stata stop running the do-file wherever it is placed.

Variables, global macros, local macros, and matrices can be used for storing and referring to data and these are made use of extensively in programs. For example, we may wish to subtract the mean of x from x. Interactively, we would use

```
summarize x
```

to find out what the mean value is and then subtract that value from x. However, we should not type the value of the mean into a do-file because the result would no longer be valid if the data change. Instead, we can access the mean computed by `summarize` using `r(mean)`:

```
quietly summarize x
gen xnew=x-r(mean)
```

Most Stata commands are *r class*, meaning that they store results that may be accessed using `r()` with the appropriate name inside the brackets. Estimation commands store the results in `e()`. In order to find out under what names constants are stored, see the "Stored Results" section for the command of interest in the *Stata Reference Manuals*.

If a local macro is defined without using the = sign, anything can appear on the right-hand side and typing the local macro name in single quotes has the same effect as typing whatever appeared on the right-hand side in the definition of the macro. For example, if we have a variable y, we can use the commands

```
local a y
disp "`a'[1] = " `a'[1]
```

```
y[1] = 4.6169958
```

Local macros are only 'visible' within the do-file or program in which they are defined. Global macros may be defined using

```
global a=1
```

and accessed by prefixing them with a dollar, for example,

```
gen b=$a
```

Sometimes it is useful to have a general set of commands (or a program) that may be applied in different situations. It is then essential that variable names and parameters specific to the application can be passed to the program. If the commands are stored in a do-file, the 'arguments' with which the do-file will be used, are referred to as `1´, `2´ etc inside the do-file. For example, a do-file containing the command

```
list `1´ `2´
```

may be run using

```
do filename x1 x2
```

to cause x1 and x2 to be listed. Alternatively, we can define a program which can be called without using the do command in much the same way as Stata's own commands. This is done by enclosing the set of commands by

```
program define progname
end
```

After running the program definition, we can run the program by typing the program name and arguments.

A frequently used programming tool both for use in do-files and in program definitions is while. For example, below we define a program called mylist that lists the first three observations of each variable in a variable list:

```
program define mylist
while "`1´"~=""{   /* outer loop: loop through variables */
      local x `1´
      local i=1
      display "`x´"
      while `i´ <=3   /* inner loop: loop through observations */
           display `x´[`i´]
           local i=`i´+1   /* next observation */
      }
      mac shift   /* next variable */
      display " "

   end
```

We can run the program using the command

```
mylist x y z
```

The inner loop simply displays the `i´th element of the variable `x´ for `i´ from 1 to 3. The outer loop uses the macro `1´ as follows: At the beginning, the macros `1´, `2´ and `3´ contain the arguments x, y, and z, respectively, with which mylist was called. The command

```
mac shift
```

shifts the contents of `2´ into `1´ and those of `3´ into `2´, etc. Therefore, the outer loop steps through variables x, y, and z.

A program may be defined by typing it into the Commands window. This is almost never done in practice; however, a more useful method is to define the program within a do-file where it can easily be edited. Note that once the program has been loaded into memory (by running the program **define** commands), it has to be cleared from memory using **program drop** before it can be redefined. It is therefore useful to have the command

```
capture program drop mylist
```

in the do-file before the **program define** command, where **capture** ensures that the do-file continues running even if **mylist** does not yet exist.

A program may also be saved in a separate file (containing only the program definition) of the same name as the program itself and having the extension *.ado*. If the ado-file (automatic do-file) is in a directory in which Stata looks for ado-files, for example the current directory, it can be executed simply by typing the name of the file. There is no need to load the program first (by running the program definition). To find out where Stata looks for ado-files, type

```
adopath
```

This lists various directories including \ado\personal/, the directory where personal ado-files may be stored. Many of Stata's own commands are actually ado-files stored in the **ado** subdirectory of the directory where **wstata.exe** is located.

1.10 Keeping Stata up to date

Stata Corporation continually updates the current version of Stata. If the computer is connected to the Internet, Stata can be updated by issuing the command

```
update all
```

Ado-files are then downloaded and stored in the correct directory. If the executable has changed since the last update, a file *wstata.bin* is also downloaded. This file should be used to overwrite **wstata.exe** after saving the latter under a new name, e.g., **wstata.old**. The command **help whatsnew** lists all the changes since the release of the present version of Stata. In addition to Stata's official updates to the package, new user-written functions are available every two months. Descriptions of these functions are published in the *Stata Technical Bulletin* (STB) and the functions themselves may be downloaded from the Web. One way of finding out about the latest programs submitted to the

STB is to subscribe to the journal which is inexpensive. Another way is via the **search** command. For example, running

```
search meta
```

gives a long list of entries including one on STB-42

```
STB-42  sbe16.1  . . . . . New syntax and output for the meta-analysis command
        (help meta if installed) . . . . . . . . . . .  S. Sharp and J. Sterne
        3/98    STB Reprints Vol 7, pages 106--108
```

which reveals that STB-42 has a directory in it called sbe16.1 containing files for "New syntax and output for the meta-analysis command" and that help on the new command may be found using **help meta**, but only after the program has been installed.

An easy way of downloading this program is to click on **Help - STB & User-written Programs**, select **http://www.stata.com**, click on **stb**, then **stb42**, **sbe16_1** and finally, click on **(click here to install)**.

The program can also be installed using the commands

```
net from http://www.stata.com
net cd stb
net cd stb42
net install sbe16_1
```

Not all user-defined programs are included in any STB (yet). Other ado-files may be found on the Statalist archive under "Contributed ADO Files" (direct address http://ideas.uqam.ca/ideas/data/bocbocode.html).

1.11 Exercises

1. Use an editor (e.g., Notepad, PFE or a word-processor) to generate the dataset *test.dat* given below, where the columns are separated by tabs (make sure to save it as a text only, or ASCII, file).

v1	v2	v3
1	3	5
2	16	3
5	12	2

2. Read the data into Stata using **insheet** (see **help insheet**).

3. Click into the Data Editor and type in the variable **sex** with values 1 2 and 1.

4. Define value labels for sex (1=male, 2=female).

5. Use **gen** to generate **id**, a subject index (from 1 to 3).

6. Use **rename** to rename the variables **v1** to **v3** to **time1** to **time3**. Aso try doing this in a single command using **for**.

7. Use **reshape** to convert the dataset to long shape.

8. Generate a variable **d** that is equal to the squared difference between the variable **time** at each occasion and the average of **time** for each subject.

9. Drop the observation corresponding to the third occasion for **id**=2.

CHAPTER 2

Data Description and Simple Inference: Female Psychiatric Patients

2.1 Description of data

The data to be used in this chapter consist of observations on 8 variables for 118 female psychiatric patients and are available in Hand et al. (1994). The variables are as follows:

- age: age in years
- IQ: intelligence questionnaire score
- anxiety: anxiety (1=none, 2=mild, 3=moderate, 4=severe)
- depress: depression (1=none, 2=mild, 3=moderate, 4=severe)
- sleep: can you sleep normally? (1=yes, 2=no)
- sex: have you lost interest in sex? (1=no, 2=yes)
- life: have you thought recently about ending your life? (1=no, 2=yes)
- weight: weight change over last six months (in lb)

The data are given in Table 2.1 with missing values coded as -99. One question of interest is how the women who have recently thought about ending their lives differ from those who have not. Also of interest are the correlations between anxiety and depression and between weight change, age, and IQ.

Table 2.1: Data in fem.dat

id	age	IQ	anx	depress	sleep	sex	life	weight
1	39	94	2	2	2	2	2	4.9
2	41	89	2	2	2	2	2	2.2
3	42	83	3	3	3	2	2	4.0
4	30	99	2	2	2	2	2	-2.6
5	35	94	2	1	1	2	1	-0.3
6	44	90	-99	1	2	1	1	0.9
7	31	94	2	2	-99	2	2	-1.5
8	39	87	3	2	2	2	1	3.5
9	35	-99	3	2	2	2	2	-1.2
10	33	92	2	2	2	2	2	0.8
11	38	92	2	1	1	1	1	-1.9
12	31	94	2	2	2	-99	1	5.5

Table 2.1: Data in fem.dat (continued)

13	40	91	3	2	2	2	1	2.7
14	44	86	2	2	2	2	2	4.4
15	43	90	3	2	2	2	2	3.2
16	32	-99	1	1	1	2	1	-1.5
17	32	91	1	2	2	-99	1	-1.9
18	43	82	4	3	2	2	2	8.3
19	46	86	3	2	2	2	2	3.6
20	30	88	2	2	2	2	1	1.4
21	34	97	3	3	-99	2	2	-99.0
22	37	96	3	2	2	2	1	-99.0
23	35	95	2	1	2	2	1	-1.0
24	45	87	2	2	2	2	2	6.5
25	35	103	2	2	2	2	1	-2.1
26	31	-99	2	2	2	2	1	-0.4
27	32	91	2	2	2	2	1	-1.9
28	44	87	2	2	2	2	2	3.7
29	40	91	3	3	2	2	2	4.5
30	42	89	3	3	2	2	2	4.2
31	36	92	3	-99	2	2	2	-99.0
32	42	84	3	3	2	2	2	1.7
33	46	94	2	-99	2	2	2	4.8
34	41	92	2	1	2	2	1	1.7
35	30	96	-99	2	2	2	2	-3.0
36	39	96	2	2	2	1	1	0.8
37	40	86	2	3	2	2	2	1.5
38	42	92	3	2	2	2	1	1.3
39	35	102	2	2	2	2	2	3.0
40	31	82	2	2	2	2	1	1.0
41	33	92	3	3	2	2	2	1.5
42	43	90	-99	-99	2	2	2	3.4
43	37	92	2	1	1	1	1	-99.0
44	32	88	4	2	2	2	1	-99.0
45	34	98	2	2	2	2	-99	0.6
46	34	93	3	2	2	2	2	0.6
47	42	90	2	1	1	2	1	3.3
48	41	91	2	1	1	1	1	4.8
49	31	-99	3	1	2	2	1	-2.2
50	32	92	3	2	2	2	2	1.0
51	29	92	2	2	2	1	2	-1.2
52	41	91	2	2	2	2	2	4.0
53	39	91	2	2	2	2	2	5.9
54	41	86	2	1	1	2	1	0.2
55	34	95	2	1	1	2	1	3.5
56	39	91	1	1	2	1	1	2.9
57	35	96	3	2	2	1	1	-0.6

Table 2.1: Data in fem.dat (continued)

58	31	100	2	2	2	2	2	-0.6
59	32	99	4	3	2	2	2	-2.5
60	41	89	2	1	2	1	1	3.2
61	41	89	3	2	2	2	2	2.1
62	44	98	3	2	2	2	2	3.8
63	35	98	2	2	2	2	1	-2.4
64	41	103	2	2	2	2	2	-0.8
65	41	91	3	1	2	2	1	5.8
66	42	91	4	3	-99	-99	2	2.5
67	33	94	2	2	2	2	1	-1.8
68	41	91	2	1	2	2	1	4.3
69	43	85	2	2	2	1	1	-99.0
70	37	92	1	1	2	2	1	1.0
71	36	96	3	3	2	2	2	3.5
72	44	90	2	-99	2	2	2	3.3
73	42	87	2	2	2	1	2	-0.7
74	31	95	2	3	2	2	2	-1.6
75	29	95	3	3	2	2	2	-0.2
76	32	87	1	1	2	2	1	-3.7
77	35	95	2	2	2	2	2	3.8
78	42	88	1	1	1	2	1	-1.0
79	32	94	2	2	2	2	1	4.7
80	39	-99	3	2	2	2	2	-4.9
81	34	-99	3	-99	2	2	1	-99.0
82	34	87	3	3	2	2	1	2.2
83	42	92	1	1	2	1	1	5.0
84	43	86	2	3	2	2	2	0.4
85	31	93	-99	2	2	2	2	-4.2
86	31	92	2	2	2	2	1	-1.1
87	36	106	2	2	2	1	2	-1.0
88	37	93	2	2	2	2	2	4.2
89	43	95	2	2	2	2	1	2.4
90	32	95	3	2	2	2	2	4.9
91	32	92	-99	-99	-99	2	2	3.0
92	32	98	2	2	2	2	2	-0.3
93	43	92	2	2	2	2	2	1.2
94	41	88	2	2	2	2	1	2.6
95	43	85	1	1	2	2	1	1.9
96	39	92	2	2	2	2	1	3.5
97	41	84	2	2	2	2	2	-0.6
98	41	92	2	1	2	2	1	1.4
99	32	91	2	2	2	2	2	5.7
100	44	86	3	2	2	2	2	4.6
101	42	92	3	2	2	2	1	-99.0
102	39	89	2	2	2	2	1	2.0

Table 2.1: Data in fem.dat (continued)

103	45	-99	2	2	2	2	2	0.6
104	39	96	3	-99	2	2	2	-99.0
105	31	97	2	-99	-99	-99	2	2.8
106	34	92	3	2	2	2	2	-2.1
107	41	92	2	2	2	2	2	-2.5
108	33	98	3	2	2	2	2	2.5
109	34	91	2	1	1	2	1	5.7
110	42	91	3	3	2	2	2	2.4
111	40	89	3	1	1	1	1	1.5
112	35	94	3	3	2	2	2	1.7
113	41	90	3	2	2	2	2	2.5
114	32	96	2	1	1	2	1	-99.0
115	39	87	2	2	2	1	2	-99.0
116	41	86	3	2	1	1	2	-1.0
117	33	89	1	1	1	1	1	6.5
118	42	-99	3	2	2	2	2	4.9

2.2 Group comparison and correlations

We have interval scale variables (weight change, age, and IQ), ordinal variables (anxiety and depression), and categorical, dichotomous variables (sex and sleep) that we wish to compare between two groups of women: those who have thought about ending their lives and those who have not.

For interval scale variables, the most common statistical test is the t-test which assumes that the observations in the two groups are independent and are sampled from two populations each having a normal distribution and equal variances. A nonparametric alternative (which does not rely on the latter two assumptions) is the Mann-Whitney U-test.

For ordinal variables, either the Mann-Whitney U-test or a χ^2-test may be appropriate depending on the number of levels of the ordinal variable. The latter test can also be used to compare dichotomous variables between the groups.

Continuous variables can be correlated using the Pearson correlation. If we are interested in the question whether the correlations differ significantly from zero, then a hypothesis test is available that assumes bivariate normality. A significance test not making this distributional assumption is also available; it is based on the correlation of the ranked variables, the Spearman rank correlation. Finally, if variables have only few categories, Kendall's tau-b provides a useful measure of correlation (see, e.g., Sprent, 1993).

2.3 Analysis using Stata

Assuming the data have been saved from a spread-sheet or statistical package (for example SPSS) as a tab-delimited ASCII file, *fem.dat*, they can be read using the instruction

```
insheet using fem.dat, clear
```

There are missing values which have been coded as -99. We replace these with Stata's missing value code '.' using

```
mvdecode _all, mv(-99)
```

The variable `sleep` has been entered incorrectly for subject 3. Such data entry errors can be detected using the command

```
codebook
```

which displays information on all variables; the output for `sleep` is shown below:

```
sleep ------------------------------------------------------------ SLEEP
                        type:  numeric (byte)

                       range:  [1,3]                       units:  1
               unique values:  3                   coded missing:  5 / 118

                  tabulation:  Freq.  Value
                                  14  1
                                  98  2
                                   1  3
```

Alternatively, we can detect errors using the `assert` command. For `sleep`, we would type

```
assert sleep==1|sleep==2|sleep==.
```

```
1 contradiction out of 118
assertion is false
```

Since we do not know what the correct code for `sleep` should have been, we can replace the incorrect value of 3 by 'missing'

```
replace sleep=. if sleep==3
```

In order to have consistent coding for "yes" and "no", we recode the variable `sleep`

```
recode sleep 1=2 2=1
```

and to avoid confusion in the future, we label the values as follows:

```
label define yn 1 no 2 yes
label values sex yn
label values life yn
label values sleep yn
```

The last three commands could also have been carried out in a **for** loop

```
for var sex life sleep: label values X yn
```

First, we can compare the groups who have and have not thought about ending their lives by tabulating summary statistics of various variables for the two groups. For example, for **IQ**, we type

```
table life, contents(mean iq sd iq)
```

```
----------+----------------------
  LIFE |    mean(iq)        sd(iq)
----------+----------------------
    no |    91.27084      3.757204
   yes |    92.09836        5.0223
----------+----------------------
```

To assess whether the groups appear to differ in their weight loss over the past six months and to informally check assumptions for an independent samples t-test, we plot the variable **weight** as a boxplot for each group after defining appropriate labels:

```
label variable weight "weight change in last six months"
sort life
graph weight, box by(life) ylabel yline(0) gap(5) /*
     */ b2("have you recently thought about ending your life?")
```

giving the graph shown in Figure 2.1. The **yline(0)** option has placed a horizontal line at 0. (Note that in the instructions above, the "brackets" for comments, **/*** and ***/** were used to make Stata ignore the line breaks in the middle of the **graph** command in a d-file). The groups do not seem to differ much in their median weight change and the assumptions for the t-test seem reasonable because the distributions are symmetric with similar spread.

We can also check the assumption of normality more formally by plotting a normal quantile plot of suitably defined residuals. Here the difference between the observed weight changes and the group-specific mean weight changes can be used. If the normality assumption is satisfied, the quantiles of the residuals should be linearly related to the quantiles of the normal distribution. The residuals can be computed and plotted using

```
egen res=mean(weight), by(life)
replace res=weight-res
label variable res "residuals of t-test for weight"
qnorm res, gap(5) xlab ylab t1("normal q-q plot")
```

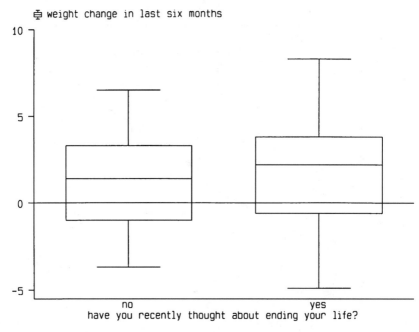

Figure 2.1 *Box-plot of weight by group.*

where gap(5) was used to reduce the gap between the vertical axis and the axis title (the default gap is 8). The points in the Q-Q plot in Figure 2.2 are sufficiently close to the staight line.

We could also test whether the variances differ significantly using

```
sdtest weight, by(life)
```

giving the output shown in Display 2.1. which shows that there is no significant difference (p=0.57) between the variances. Note that the test for equal variances is only appropriate if the variable may be assumed to be normally distributed in each population. Having checked the assumptions, we carry out a t-test for weight change:

```
ttest weight, by(life)
```

Display 2.2 shows that the two-tailed significance is p=0.55 and the 95% confidence interval for the mean difference in weight change between those who have thought about ending their lives and those who have not is from -0.74 pounds to 1.39 pounds. Therefore, there is no evidence that the populations differ in their mean weight change, but we also cannot rule out a difference in mean weight change of as much as 1.4 pounds in 6 months.

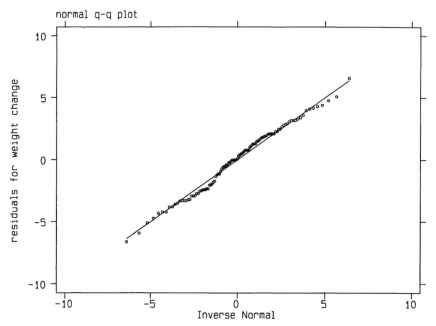

Figure 2.2 *Normal Q-Q plot of residuals of weight change.*

```
Variance ratio test

-----------------------------------------------------------------------
   Group |     Obs        Mean    Std. Err.    Std. Dev.   [95% Conf. Interval]
---------+-------------------------------------------------------------
      no |      45    1.408889    .3889616     2.609234     .6249883     2.19279
     yes |      61    1.731148    .3617847     2.825629     1.00747    2.454825
---------+-------------------------------------------------------------
combined |     106     1.59434    .2649478     2.727805    1.068997    2.119682
-----------------------------------------------------------------------

                        Ho: sd(no) = sd(yes)

            F(44,60) observed   = F_obs         =      0.853
            F(44,60) lower tail = F_L   = F_obs =      0.853
            F(44,60) upper tail = F_U   = 1/F_obs =    1.173

Ha: sd(no) < sd(yes)        Ha: sd(no) ~= sd(yes)        Ha: sd(no) > sd(yes)
 P < F_obs = 0.2919        P < F_L + P > F_U = 0.5724      P > F_obs = 0.7081
```

Display 2.1

```
Two-sample t test with equal variances

-------------------------------------------------------------------------------
   Group |    Obs        Mean    Std. Err.    Std. Dev.   [95% Conf. Interval]
---------+---------------------------------------------------------------------
      no |     45    1.408889    .3889616     2.609234    .6249883     2.19279
     yes |     61    1.731148    .3617847     2.825629    1.00747     2.454825
---------+---------------------------------------------------------------------
combined |    106     1.59434    .2649478     2.727805    1.068997    2.119682
---------+---------------------------------------------------------------------
    diff |            -.3222587   .5376805                 -1.388499    .743982
-------------------------------------------------------------------------------
Degrees of freedom: 104

                      Ho: mean(no) - mean(yes) = diff = 0

    Ha: diff < 0                 Ha: diff ~= 0                 Ha: diff > 0
      t =  -0.5993                 t =  -0.5993                 t =  -0.5993
    P < t =   0.2751             P > |t| =   0.5502           P > t =   0.7249
```

Display 2.2

We can use a χ^2- test to test for differences in depression between the two groups and display the corresponding cross-tabulation together with the percentage of women in each category of depression by group using the single command

```
tabulate life depress, row chi2
```

```
         | DEPRESS
    LIFE |        1          2          3 |     Total
---------+---------------------------------+----------
      no |       26         24          1 |        51
         |    50.98      47.06       1.96 |    100.00
---------+---------------------------------+----------
     yes |        0         42         16 |        58
         |     0.00      72.41      27.59 |    100.00
---------+---------------------------------+----------
   Total |       26         66         17 |       109
         |    23.85      60.55      15.60 |    100.00

        Pearson chi2(2) =   43.8758   Pr = 0.000
```

There is a significant association between **depress** and **life** with none of the subjects who have thought about ending their lives having zero depression compared with 51% of those who have not. Note that this test does not take account of the ordinal nature of depression and is therefore likely to be less sensitive than, for example, ordinal regression (see Chapter 6). Fisher's exact test can be obtained without having to reproduce the table as follows:

```
tabulate life depress, exact nofreq
```

```
              Fisher's exact =                 0.000
```

Similarly, for sex, we can obtain the table and both the χ^2-test and Fisher's exact tests using

```
tabulate life sex, row chi2 exact
```

```
            | SEX
      LIFE  |        no        yes  |     Total
    --------+----------------------+----------
        no  |        12         38  |        50
            |     24.00      76.00  |    100.00
    --------+----------------------+----------
       yes  |         5         58  |        63
            |      7.94      92.06  |    100.00
    --------+----------------------+----------
     Total  |        17         96  |       113
            |     15.04      84.96  |    100.00

              Pearson chi2(1) =    5.6279   Pr = 0.018
              Fisher's exact  =                 0.032
       1-sided Fisher's exact =                 0.017
```

Therefore, those who have thought about ending their lives are more likely to have lost interest in sex than those who have not (92% compared with 76%). We can explore correlations between the three variables weight, IQ, and age using a single command

```
corr weight iq age
```

```
(obs=100)

            |   weight        iq       age
    --------+---------------------------------
    weight  |   1.0000
        iq  |  -0.2920    1.0000
       age  |   0.4131   -0.4363    1.0000
```

The correlation matrix has been evaluated for those 100 observations that had complete data on all three variables. The command pwcorr may be used to include, for each correlation, all observations that have complete data for the corresponding pair of variables, resulting in different sample sizes for different correlations. The sample sizes and p-values can be displayed simultaneously as follows:

```
pwcorr weight iq age, obs sig
```

```
              |   weight        iq       age
   -----------+--------------------------------
       weight |   1.0000
              |
              |      107
              |
           iq |  -0.2920    1.0000
              |   0.0032
              |      100       110
              |
          age |   0.4156   -0.4345    1.0000
              |   0.0000    0.0000
              |      107       110       118
              |
```

The corresponding scatter-plot matrix is obtained using

```
graph weight iq age, matrix half jitter(1)
```

where `jitter(1)` randomly moves the points by a very small amount to stop them overlapping completely due to the discrete nature of age and IQ. The resulting graph is shown in Figure 2.3. Thus, older psychiatric females tend to

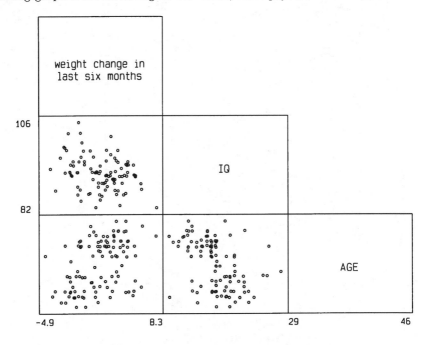

Figure 2.3 *Scatter-plot matrix for weight, IQ, and age.*

put on more weight than younger ones as do less intelligent women. However, older women in this sample also tended to be less intelligent so that age and intelligence are confounded.

It would be interesting to see whether those who have thought about ending their lives have the same relationship between age and weight change as do those who have not. In order to form a single scatter-plot with different symbols representing the two groups, we must use a single variable for the x-axis (`age`) and plot two separate variables `wgt1` and `wgt2` which contain the weight changes for groups 1 and 2, respectively:

```
gen wgt1 = weight if life==2
gen wgt2 = weight if life==1
label variable wgt1 "no"
label variable wgt2 "yes"
graph wgt1 wgt2 age, s(dp) xlabel ylabel        /*
       */ 11("weight change in last 6 months")  /*
```

The resulting graph in Figure 2.4 shows that within both groups, higher age is associated with larger weight increases and the groups do not form distinct clusters.

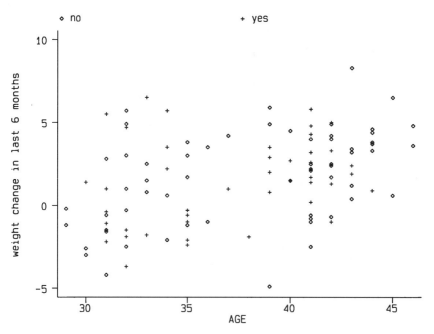

Figure 2.4 *Scatter-plot of weight against age.*

Finally, an appropriate correlation between depression and anxiety is Kendall's tau-b which can be obtained using

```
ktau depress anxiety
```

```
Number of obs =        107
Kendall's tau-a =          0.2827
Kendall's tau-b =          0.4951
Kendall's score =       1603
     SE of score =        288.279   (corrected for ties)


Test of Ho: depress and anxiety independent
        Pr > |z| =          0.0000  (continuity corrected)
```

giving a value of 0.50 with an approximate p-value of $p < 0.001$.

2.4 Exercises

1. Tabulate the mean weight change by level of depression.

2. Using `for`, tabulate the means and standard deviations by `life` for each of the variables `age`, `iq` and `weight`.

3. Use `search nonparametric` or `search mann` or `search whitney` to find help on how to run the Mann-Whitney U-test.

4. Compare the weight changes between the two groups using the Mann Whitney U-test.

5. Form a scatter-plot for `IQ` and `age` using different symbols for the two groups (life=1 and life=2). Explore the use of the option `jitter(#)` for different integers `#` to stop symbols overlapping.

6. Having tried out all these commands interactively, create a do-file containing these commands and run the do-file. In the `graph` commands, use the option `saving(filename,replace)` to save the graphs in the current directory and view the graphs later using the command `graph using filename`.

See also Exercises in Chapter 6.

CHAPTER 3

Multiple Regression: Determinants of Pollution in U.S. Cities

3.1 Description of data

Data on air pollution in 41 U.S. cities were collected by Sokal and Rohlf (1981) from several U.S. government publications and are reproduced here in Table 3.1. (The data are also available in Hand et al. (1994).) There is a single dependent variable, so2, the annual mean concentration of sulphur dioxide, in micrograms per cubic meter. These data generally relate to means for the three years 1969 to 1971. The values of six explanatory variables, two of which concern human ecology and four climate, are also recorded; details are as follows:

- temp: average annual temperature in °F

- manuf: number of manufacturing enterprises employing 20 or more workers

- pop: population size (1970 census) in thousands

- wind: average annual wind speed in miles per hour

- precip: average annual precipitation in inches

- days: average number of days with precipitation per year.

The main question of interest about this data is how the pollution level as measured by sulphur dioxide concentration is determined by the six explanatory variables. The central method of analysis will be *multiple regression*.

3.2 The multiple regression model

The multiple regression model has the general form

$$y_i = \beta_0 + \beta_1 x_{1i} + \beta_2 x_{2i} + \cdots + \beta_p x_{pi} + \epsilon_i \tag{3.1}$$

where y is a continuous response variable, $x_1, x_2, \cdots, x_p$ are a set of explanatory variables and ϵ is a residual term. The regression coefficients, $\beta_0, \beta_1, \cdots, \beta_p$ are generally estimated by least squares. Significance tests for the regression coefficients can be derived by assuming that the residual terms are normally distributed with zero mean and constant variance σ^2.

For n observations of the response and explanatory variables, the regression

Table 3.1: Data in usair.dat

Town	SO2	temp	manuf	pop	wind	precip	days
Phoenix	10	70.3	213	582	6.0	7.05	36
Little Rock	13	61.0	91	132	8.2	48.52	100
San Francisco	12	56.7	453	716	8.7	20.66	67
Denver	17	51.9	454	515	9.0	12.95	86
Hartford	56	49.1	412	158	9.0	43.37	127
Wilmington	36	54.0	80	80	9.0	40.25	114
Washington	29	57.3	434	757	9.3	38.89	111
Jackson	14	68.4	136	529	8.8	54.47	116
Miami	10	75.5	207	335	9.0	59.80	128
Atlanta	24	61.5	368	497	9.1	48.34	115
Chicago	110	50.6	3344	3369	10.4	34.44	122
Indiana	28	52.3	361	746	9.7	38.74	121
Des Moines	17	49.0	104	201	11.2	30.85	103
Wichita	8	56.6	125	277	12.7	30.58	82
Louisville	30	55.6	291	593	8.3	43.11	123
New Orleans	9	68.3	204	361	8.4	56.77	113
Baltimore	47	55.0	625	905	9.6	41.31	111
Detroit	35	49.9	1064	1513	10.1	30.96	129
Minnisota	29	43.5	699	744	10.6	25.94	137
Kansas	14	54.5	381	507	10.0	37.00	99
St. Louis	56	55.9	775	622	9.5	35.89	105
Omaha	14	51.5	181	347	10.9	30.18	98
Albuquerque	11	56.8	46	244	8.9	7.77	58
Albany	46	47.6	44	116	8.8	33.36	135
Buffalo	11	47.1	391	463	12.4	36.11	166
Cincinnati	23	54.0	462	453	7.1	39.04	132
Cleveland	65	49.7	1007	751	10.9	34.99	155
Columbia	26	51.5	266	540	8.6	37.01	134
Philadelphia	69	54.6	1692	1950	9.6	39.93	115
Pittsburgh	61	50.4	347	520	9.4	36.22	147
Providence	94	50.0	343	179	10.6	42.75	125
Memphis	10	61.6	337	624	9.2	49.10	105
Nashville	18	59.4	275	448	7.9	46.00	119
Dallas	9	66.2	641	844	10.9	35.94	78
Houston	10	68.9	721	1233	10.8	48.19	103
Salt Lake City	28	51.0	137	176	8.7	15.17	89
Norfolk	31	59.3	96	308	10.6	44.68	116
Richmond	26	57.8	197	299	7.6	42.59	115
Seattle	29	51.1	379	531	9.4	38.79	164
Charleston	31	55.2	35	71	6.5	40.75	148
Milwaukee	16	45.7	569	717	11.8	29.07	123

model may be written concisely as

$$\mathbf{y} = \mathbf{X}\boldsymbol{\beta} + \boldsymbol{\epsilon} \tag{3.2}$$

where $\mathbf{y}$ is the $n \times 1$ vector of responses, $\mathbf{X}$ is an $n \times (p+1)$ matrix of known constants, the first column containing a series of ones corresponding to the term β_0 in (3.1) and the remaining columns values of the explanatory variables. The elements of the vector $\boldsymbol{\beta}$ are the regression coefficients $\beta_0, \cdots, \beta_p$, and those of the vector $\boldsymbol{\epsilon}$, the residual terms $\epsilon_1, \cdots, \epsilon_n$. For full details of multiple regression see, for example, Rawlings (1988).

3.3 Analysis using Stata

Assuming the data are available as an ASCII file *usair.dat* in the current directory and that the file contains city names as given in Table 3.1, they may be read in for analysis using the following instruction:

```
infile str10 town so2 temp manuf pop /*
    */ wind precip days using usair.dat
```

Before undertaking a formal regression analysis of these data, it will be helpful to examine them graphically using a scatter-plot matrix. Such a display is useful in assessing the general relationships between the variables, in identifying possible outliers, and in highlighting potential collinearity problems amongst the explanatory variables. The basic plot can be obtained using

```
graph so2 temp manuf pop wind precip days, matrix
```

The resulting diagram is shown in Figure 3.1. Several of the scatter-plots show evidence of outliers and the relationship between **manuf** and **pop** is very strong suggesting that using both as explanatory variables in a regression analysis may lead to problems (see later). The relationships of particular interest, namely those between **so2** and the explanatory variables (the relevant scatterplots are those in the first row of Figure 3.1), indicate some possible nonlinearity. A more informative, although slightly more 'messy' diagram can be obtained if the plotted points are labeled with the associated town name. The necessary Stata instruction is

```
graph so2-days, matrix symbol([town]) tr(3) ps(150)
```

The **symbol()** option labels the points with the names in the **town** variable; if, however, the full name were used, the diagram would be very difficult to read. Consequently the **trim()** option, abbreviated **tr()**, is used to select the first three characters of each name for plotting and the **psize()** option, abbreviated **ps()**, is used to increase the size of these characters to 150% compared with the usual 100% size. The resulting diagram appears in Figure 3.2. Clearly, Chicago and to a lesser extent Philadelphia might be considered outliers. Chicago has

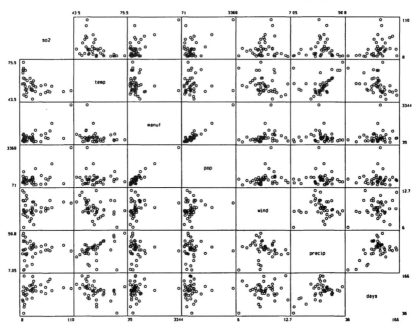

Figure 3.1 *Scatter-plot matrix.*

such a high degree of pollution compared with the other cities that it should perhaps be considered as a special case and excluded from further analysis. A new data file with Chicago removed can be generated using

```
clear
infile str10 town so2 temp manuf pop wind /*
      */ precip days using usair.dat if town~="Chicago"
```

or by dropping the opservation using

```
drop if town=="Chicago"
```

The command **regress** may be used to fit a basic multiple regression model. The necessary Stata instruction to regress sulphur dioxide concentration on the six explanatory variables is

```
regress so2 temp manuf pop wind precip days
```

or, alternatively,

```
regress so2 temp-days
```

(see Display 3.1)

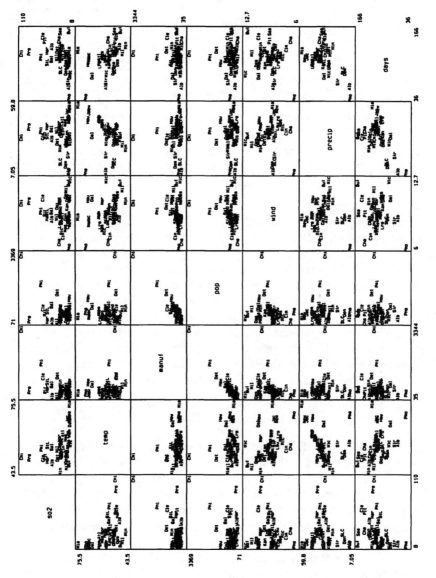

Figure 3.2 *Scatter-plot matrix with town labels.*

```
  Source |       SS       df       MS                 Number of obs =       40
---------+------------------------------              F(  6,    33) =     6.20
   Model |   8203.60523    6  1367.26754              Prob > F      =   0.0002
Residual |   7282.29477   33  220.675599              R-squared     =   0.5297
---------+------------------------------              Adj R-squared =   0.4442
   Total |    15485.90    39  397.074359              Root MSE      =   14.855

---------------------------------------------------------------------------------
     so2 |     Coef.    Std. Err.       t      P>|t|       [95% Conf. Interval]
---------+-----------------------------------------------------------------------
    temp |  -1.268452    .6305259    -2.012    0.052     -2.551266     .0143631
   manuf |   .0654927    .0181777     3.603    0.001      .0285098     .1024756
     pop |   -.039431    .0155342    -2.538    0.016     -.0710357    -.0078264
    wind |  -3.198267    1.859713    -1.720    0.095     -6.981881     .5853469
  precip |   .5136846    .3687273     1.393    0.173     -.2364966    1.263866
    days |  -.0532051    .1653576    -0.322    0.750     -.3896277     .2832175
   _cons |   111.8709    48.07439     2.327    0.026      14.06278     209.679
---------------------------------------------------------------------------------
```

Display 3.1

The main features of interest are the analysis of variance table and the parameter estimates. In the former, the ratio of the model mean square to the residual mean square gives an F-test for the hypothesis that *all* the regression coefficients in the fitted model are zero. The resulting F-statistic with 6 and 33 degrees of freedom takes the value 6.20 and is shown on the right-hand side; the associated P-value is very small. Consequently, the hypothesis is rejected. The square of the multiple correlation coefficient (R^2) is 0.53 showing that 53% of the variance of sulphur dioxide concentration is accounted for by the six explanatory variables of interest. The adjusted R^2 statistic is an estimate of the population R^2 taking account of the fact that the parameters were estimated from the data. The statistic is calculated as

$$\text{adj } R^2 = 1 - \frac{(n - i)(1 - R^2)}{n - p} \tag{3.3}$$

where n is the number of observations used in fitting the model, and i is an indicator variable that takes the value 1 if the model includes an intercept and 0 otherwise. The root MSE is simply the square root of the residual mean square in the analysis of variance table, which itself is an estimate of the parameter σ^2. The estimated regression coefficients give the estimated change in the response variable produced by a unit change in the corresponding explanatory variable with the remaining explanatory variables held constant.

One concern generated by the initial graphical material on this data was the strong relationship between the two explanatory variables **manuf** and **pop**. The correlation of these two variables is obtained by using

```
correlate manuf pop
```

```
(obs=40)

          |    manuf        pop
----------+-------------------
   manuf|   1.0000
     pop|   0.8906     1.0000
```

The strong linear dependence might be a source of collinearity problems and can be investigated further by calculating what are known as *variance inflation factors* for each of the explanatory variables. These are given by

$$\text{VIF}(x_i) = \frac{1}{1 - R_i^2} \tag{3.4}$$

where $\text{VIF}(x_i)$ is the variance inflation factor for explanatory variable x_i and R_i^2 is the square of the multiple correlation coefficient obtained from regressing x_i on the remaining explanatory variables.

The variance inflation factors can be found in Stata by following **regress** with **vif**:

vif

```
Variable |      VIF      1/VIF
---------+------------------------
   manuf |     6.28     0.159275
     pop |     6.13     0.163165
    temp |     3.72     0.269156
    days |     3.47     0.287862
  precip |     3.41     0.293125
    wind |     1.26     0.790619
---------+------------------------
Mean VIF |     4.05
```

Chatterjee and Price (1991) give the following 'rules-of-thumb' for evaluating these factors:

- Values larger than 10 give evidence of multicollinearity.

- A mean of the factors considerably larger than one suggests multicollinearity.

Here there are no values greater than 10 (as an exercise we suggest readers also calculate the VIFs when the observations for Chicago are included), but the mean value of 4.05 gives some cause for concern. A simple (although not necessarily the best) way to proceed is to drop one of **manuf** or **pop**. We shall exclude the former and repeat the regression analysis using the five remaining explanatory variables:

regress so2 temp pop wind precip days

The output is shown in Display 3.2.

```
 Source |       SS         df       MS                    Number of obs =        40
--------+----------------------------------              F(  5,    34) =      3.58
  Model |  5339.03465       5   1067.80693               Prob > F      =    0.0105
Residual|  10146.8654      34   298.437216               R-squared     =    0.3448
--------+----------------------------------              Adj R-squared =    0.2484
  Total |   15485.90       39   397.074359               Root MSE      =    17.275


-----------------------------------------------------------------------------------
    so2 |     Coef.    Std. Err.       t      P>|t|       [95% Conf. Interval]
--------+--------------------------------------------------------------------------
   temp | -1.867665    .7072827     -2.641    0.012      -3.305037     -.430294
    pop |  .0113969    .0075627      1.507    0.141      -.0039723     .0267661
   wind | -3.126429    2.16257      -1.446    0.157       -7.5213      1.268443
 precip |  .6021108    .4278489      1.407    0.168      -.2673827     1.471604
   days | -.020149     .1920012     -0.105    0.917      -.4103424     .3700445
  _cons |  135.8565    55.36797      2.454    0.019       23.33529     248.3778
-----------------------------------------------------------------------------------
```

Display 3.2

Now recompute the variance inflation factors:

vif

```
 Variable |     VIF       1/VIF
----------+--------------------
     days |    3.46     0.288750
     temp |    3.46     0.289282
   precip |    3.40     0.294429
     wind |    1.26     0.790710
      pop |    1.07     0.931015
----------+--------------------
 Mean VIF |    2.53
```

The variance inflation factors are now satisfactory.

The very general hypothesis concerning all regression coefficients mentioned previously is not usually of great interest in most applications of multiple regression since it is most unlikely that all the chosen explanatory variables will be unrelated to the response variable. The more interesting question is whether a subset of the regression coefficients are zero, implying that not all the explanatory variables are of use in determining the response variable. A preliminary assessment of the likely importance of each explanatory variable can be made using the table of estimated regression coefficients and associated statistics. Using a conventional 5% criterion, the only 'significant' coefficient is that for the variable temp. Unfortunately, this very simple approach is not in general suitable, since in most cases the explanatory variables are correlated and the t-tests will not be independent of each other. Consequently, removing a particular variable from the regression will alter both the estimated regression coefficients of the remaining variables and their standard errors. A more involved approach to identifying important subsets of explanatory variables is therefore required. A number of procedures are available.

1. Considering some *a priori* ordering or grouping of the variables generated by the substantive questions of interest.

2. Automatic selection methods, which are of the following types:

   (a) *Forward selection*: This method starts with a model containing some of the explanatory variables and then considers variables one by one for inclusion. At each step, the variable added is the one that results in the biggest increase in the model or regression sum of squares. An F-type statistic is used to judge when further additions would not represent a significant improvement in the model.

   (b) *Backward elimination*: Here variables are considered for removal from an initial model containing all the explanatory variables. At each stage, the variable chosen for exclusion is the one leading to the smallest reduction in the regression sum of squares. Again, an F-type statistic is used to judge when further exclusions would represent a significant deterioration in the model.

   (c) *Stepwise regression*: This method is essentially a combination of the previous two. The forward selection procedure is used to add variables to an existing model and, after each addition, a backward elimination step is introduced to assess whether variables entered earlier might now be removed, because they no longer contribute significantly to the model.

In the best of all possible worlds, the final model selected by the three automatic procedures would be the same. This is often the case, but it is not guaranteed. It should also be stressed that none of the automatic procedures for selecting subsets of variables are foolproof. They must be used with care and warnings such as those given in McKay and Campbell (1982a, 1982b) concerning the validity of the F-tests used to judge whether variables should be included or eliminated, noted.

In this example, begin by considering an *a priori* grouping of the five explanatory variables since one, **pop**, relates to human ecology and the remaining four to climate. To perform a forward selection procedure with the ecology variable treated as a single term (all variables being either entered or not entered based on their joint significance) and similarly the climate terms, requires the following instruction:

```
sw regress so2 (pop) (temp wind precip days), pe(0.05)
```

(see Display 3.3)

Note the grouping as required. The **pe** term specifies the significance level of the F-test for addition to the model; terms with a p-value less than the figure specified will be included. Here, only the climate variables are shown since they are jointly significant (p=0.0119) at the significance level for inclusion.

As an illustration of the automatic selection procedures, the following Stata instruction applies the backward elimination method, with explanatory vari-

```
                        begin with empty model
 p = 0.0119 <  0.0500  adding    temp wind precip days

  Source |      SS         df      MS                Number of obs =      40
---------+------------------------------            F(  4,    35) =    3.77
   Model | 4661.27545       4  1165.31886            Prob > F      =  0.0119
Residual | 10824.6246      35  309.274987            R-squared     =  0.3010
---------+------------------------------            Adj R-squared =  0.2211
   Total |   15485.90      39  397.074359            Root MSE      =  17.586


--------------------------------------------------------------------------
     so2 |    Coef.   Std. Err.      t     P>|t|    [95% Conf. Interval]
---------+----------------------------------------------------------------
    temp | -1.689848   .7099204    -2.380   0.023    -3.131063   -.2486329
    wind | -2.309449    2.13119    -1.084   0.286    -6.635996    2.017097
  precip |  .5241595   .4323535     1.212   0.234    -.3535647    1.401884
    days |  .0119373   .1942509     0.061   0.951     -.382413    .4062876
   _cons |  123.5942   55.75236     2.217   0.033     10.41091    236.7775
--------------------------------------------------------------------------
```

Display 3.3

ables whose F-values for removal have associated p-values greater than 0.2 being removed:

```
sw regress so2 temp pop wind precip days, pr(0.2)
```

(see Display 3.4)

```
                        begin with full model
 p = 0.9170 >= 0.2000  removing days

  Source |      SS         df      MS                Number of obs =      40
---------+------------------------------            F(  4,    35) =    4.60
   Model | 5335.74801       4   1333.937            Prob > F      =  0.0043
Residual | 10150.152       35  290.004343            R-squared     =  0.3446
---------+------------------------------            Adj R-squared =  0.2696
   Total |   15485.90      39  397.074359            Root MSE      =   17.03


--------------------------------------------------------------------------
     so2 |    Coef.   Std. Err.      t     P>|t|    [95% Conf. Interval]
---------+----------------------------------------------------------------
    temp | -1.810123   .4404001    -4.110   0.000    -2.704183   -.9160635
     pop |  .0113089   .0074091     1.526   0.136    -.0037323    .0263501
    wind | -3.085284   2.096471    -1.472   0.150    -7.341347    1.170778
  precip |  .5660172   .2508601     2.256   0.030     .0567441     1.07529
   _cons |  131.3386   34.32034     3.827   0.001     61.66458    201.0126
--------------------------------------------------------------------------
```

Display 3.4

With the chosen significance level, only the variable **days** is excluded.

The next stage in the analysis should be an examination of the *residuals* from the chosen model; that is, the differences between the observed and fitted values of sulphur dioxide concentration. Such a procedure is vital for assessing model assumptions, identifying any unusual features in the data, indicating outliers, and suggesting possibly simplifying transformations. The most useful ways of examining the residuals are graphical and the most commonly used plots are as follows:

- A plot of the residuals against each explanatory variable in the model. The presence of a curvilinear relationship, for example, would suggest that a higher-order term, perhaps a quadratic in the explanatory variable, should be added to the model.

- A plot of the residuals against predicted values of the response variable. If the variance of the response appears to increase with predicted value, a transformation of the response may be in order.

- A normal probability plot of the residuals—after all systematic variation has been removed from the data, the residuals should look like a sample from the normal distribution. A plot of the ordered residuals against the expected order statistics from a normal distribution provides a graphical check on this assumption.

The first two plots can be obtained after using the **regress** procedure with the **rvpplot** and **rvfplot** instructions. For example, for the model chosen by the backward selection procedure, a plot of residuals against predicted values with the first three letters of the town name used to label the points is obtained using the instruction

```
rvfplot, symbol([town]) tr(3) ps(150) xlab ylab gap(3)
```

The resulting plot is shown in Figure 3.3, and indicates a possible problem, namely the apparently increasing variance of the residuals as the fitted values increase (see also Chapter 7). Perhaps some thought needs to be given to the possible transformations of the response variable (see exercises).

Next, graphs of the residuals plotted against each of the four explanatory variables can be obtained using the following series of instructions:

```
rvpplot pop, symbol([town]) tr(3) ps(150) xlab ylab /*
    */ ll("Residuals") gap(3)
rvpplot temp, symbol([town]) tr(3) ps(150) xlab ylab /*
    */ ll("Residuals") gap(3)
rvpplot wind, symbol([town]) tr(3) ps(150) xlab ylab /*
    */ ll("Residuals") gap(3)
rvpplot precip, symbol([town]) tr(3) ps(150) xlab ylab /*
    */ ll("Residuals") gap(3)
```

The resulting graphs are shown in Figures 3.4 to 3.7. In each graph the point corresponding to the town *Providence* is somewhat distant from the bulk of the points, and the graph for **wind** has perhaps a 'hint' of a curvilinear structure.

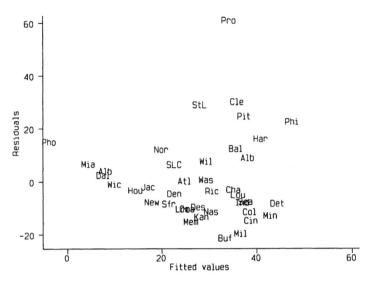

Figure 3.3 *Residuals against predicted response.*

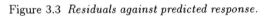

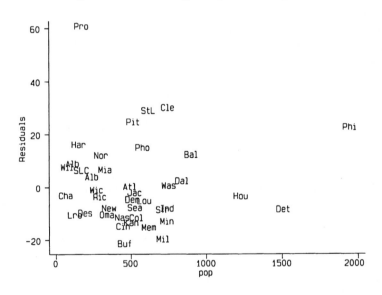

Figure 3.4 *Residuals against population.*

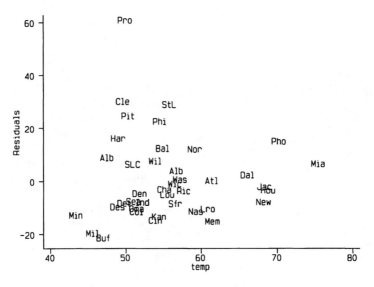

Figure 3.5 *Residuals against temperature.*

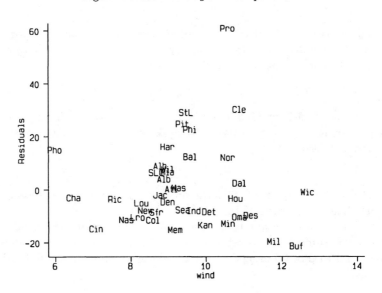

Figure 3.6 *Residuals against wind speed.*

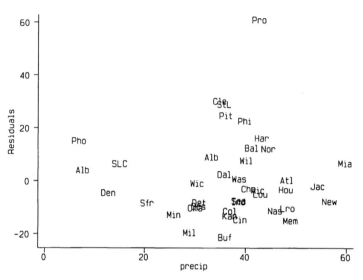

Figure 3.7 *Residuals against precipitation.*

The simple residuals plotted by `rvfplot` and `rvpplot` have a distribution that is scale dependent since the variance of each is a function of both σ^2 and the diagonal values of the so-called 'hat' matrix, $\mathbf{H}$, given by

$$\mathbf{H} = \mathbf{X}(\mathbf{X'X})^{-1}\mathbf{X'} \tag{3.5}$$

(see Cook and Weisberg (1982), for a full explanation of the hat matrix). Consequently, it is often more useful to work with a standardized version, r_i, calculated as follows:

$$r_i = \frac{y_i - \hat{y}_i}{s\sqrt{1 - h_{ii}}} \tag{3.6}$$

where s^2 is the estimate of σ^2, $\hat{y}_i$ is the predicted value of the response and h_{ii} is the ith diagonal element of $\mathbf{H}$.

These standardized residuals can be obtained by applying the `predict` instruction. For example, to obtain a normal probability plot of the standardized residuals and to plot them against the fitted values requires the following instructions:

```
predict fit
predict sdres, rstandard
pnorm sdres, gap(5)
graph sdres fit, symbol([town]) tr(3) ps(150) xlab ylab gap(3)
```

The first instruction stores the fitted values in the variable `fit`, the second stores the standardized residuals in the variable `sdres`, the third produces a

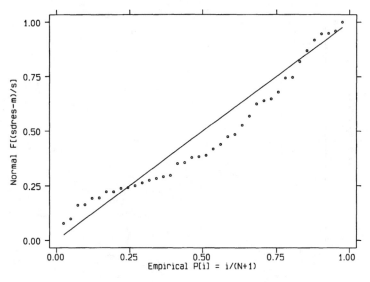

Figure 3.8 *Normal probability plot of standardized residuals.*

normal probability plot (Figure 3.8), and the last instruction produces the graph of standardized residuals against fitted values, which is shown in Figure 3.9.

The normal probability plot indicates that the distribution of the residuals departs somewhat from normality. The pattern in the plot shown in Figure 3.9 is identical to that in Figure 3.3 but here values outside (-2,2) indicate possible outliers; in this case the point corresponding to the town *Providence*. Analogous plots to those in Figures 3.4 to 3.7 could be obtained in the same way.

A rich variety of other diagnostics for investigating fitted regression models has been developed during the last decade and many of these are available with the **regress** procedure. Illustrated here is the use of two of these, namely the *partial residual plot* (Mallows (1986) and *Cook's distance statistic* (Cook, 1977, 1979). The former are useful in identifying whether, for example, quadratic or higher order terms are needed for any of the explanatory variables; the latter measures the change to the estimates of the regression coefficients that results from deleting each observation and can be used to indicate those observations that may be having an undue influence on the estimation and fitting process.

The partial residual plots are obtained using the **cprplot** command. For the four explanatory variables in the selected model for the pollution data, the required plots are obtained as follows:

```
cprplot pop, border c(k) xlab ylab gap(4) /*
```

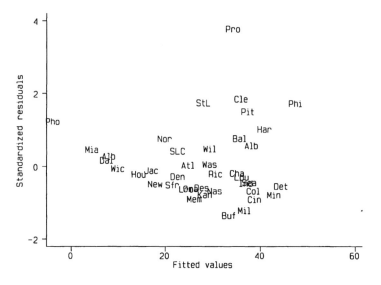

Figure 3.9 *Standardized residuals against predicted values.*

```
      */ ll("Partial Residuals")
cprplot temp, border c(k) xlab ylab gap(4) /*
      */ ll("Partial Residuals")
cprplot wind, border c(k) xlab ylab gap(4) /*
      */ ll("Partial Residuals")
cprplot precip, border c(k) xlab ylab gap(4) /*
      */ ll("Partial Residuals")
```

The two **graph** options used, **border** and **c(k)**, produce a border for the graphs and a locally weighted regression curve or *lowess*. The resulting graphs are shown in Figures 3.10 to 3.13. The graphs have to be examined for nonlinearities and whether the regression line, which has slope equal to the estimated effect of the corresponding explanatory variable in the chosen model, fits the data adequately. The added lowess curve is generally helpful for both. None of the four graphs give any obvious indication of nonlinearity.

The Cook's distance statistics are found by again using the **predict** instruction; the following calculates these statistics for the chosen model for the pollution data and lists the observations where the statistic is greater than 4/40 (4/n), which is usually the value regarded as indicating possible problems.

```
predict cook, cooksd
list town so2 cook if cook>4/40
```

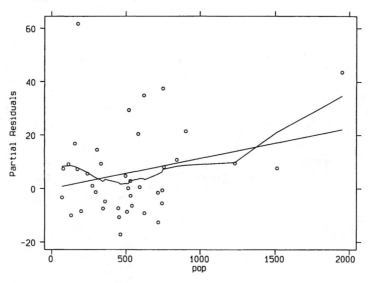

Figure 3.10 *Partial residual plot for population.*

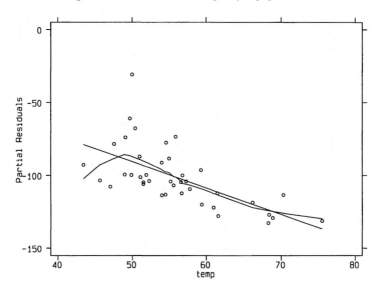

Figure 3.11 *Partial residual plot for temperature.*

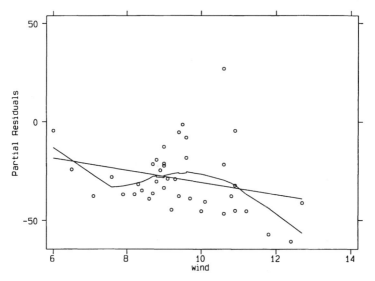

Figure 3.12 *Partial residual plot for wind speed.*

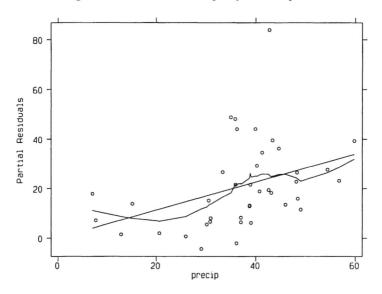

Figure 3.13 *Partial residual plot for precipitation.*

	town	so2	cook
1.	Phoenix	10	.2543286
24.	Philad	69	.3686437
29.	Provid	94	.2839324

The first instruction stores the Cook's distance statistics in the variable `cook` and the second lists details of those observations for which the statistic is above the suggested cut-off point.

There are three influential observations. Several of the diagnostic procedures used previously also suggest these observations as possibly giving rise to problems and some consideration should be given to repeating the analyses with these three observations removed in addition to the initial removal of *Chicago*.

3.4 Exercises

1. Repeat the analyses described in this chapter after removing the three possible outlying observations identified by the use of the Cook's distance statistic.

2. The solution to the high correlation of the variables `manuf` and `pop` adopted in the chapter was simply to remove the former. Investigate other possibilities such as defining a new variable `manuf/pop` in addition to `pop` to be used in the regression analysis.

3. Consider the possibility of taking a transformation of sulphur dioxide pollution before undertaking any regression analyses. For example, try a *log* transformation.

4. Explore the use of the many other diagnostic procedures available with the `regress` procedure.

See also Exercises in Chapter 12.

CHAPTER 4

Analysis of Variance I: Treating Hypertension

4.1 Description of data

Maxwell and Delaney (1990) describe a study in which the effects of 3 possible treatments for hypertension were investigated. The details of the treatments are as follows:

Treatment	Description	Levels
drug	medication	drug X, drug Y, drug Z
biofeed	physiological feedback	present, absent
diet	special diet	present, absent

All 12 combinations of the three treatments were included in a $3 \times 2 \times 2$ design. Of the subjects suffering from hypertension, 72 were recruited and 6 were allocated to each combination of treatments. Blood pressure measurements were made on each subject leading to the data shown in Table 4.1.

Table 4.1: Data in bp.raw

	Biofeedback			No Biofeedback	
drug X	drug Y	drug Z	drug X	drug Y	drug Z
		Diet absent			
170	186	180	173	189	202
175	194	187	194	194	228
165	201	199	197	217	190
180	215	170	190	206	206
160	219	204	176	199	224
158	209	194	198	195	204
		Diet present			
161	164	162	164	171	205
173	166	184	190	173	199
157	159	183	169	196	170
152	182	156	164	199	160
181	187	180	176	180	179
190	174	173	175	203	179

Questions of interest concern differences in mean blood pressure for the different levels of the three treatments and the possibility of interactions between the treatments.

4.2 Analysis of variance model

A suitable model for these data is

$$y_{ijkl} = \mu + \alpha_i + \beta_j + \gamma_k + (\alpha\beta)_{ij} + (\alpha\gamma)_{ik} + (\beta\gamma)_{jk} + (\alpha\beta\gamma)_{ijk} + \epsilon_{ijk} \quad (4.1)$$

where y_{ijkl} represents the blood pressure of the lth subject for the ith drug, the jth level of biofeedback, and the kth level of diet, μ is the overall mean, α_i, β_j and γ_k are the main effects for drugs, biofeedback and diets, $\alpha\beta, \alpha\gamma$ and $\beta\gamma$ are the first order interaction terms, $\alpha\beta\gamma$ is a second order interaction term, and ϵ_{ijkl} are the residual or error terms assumed to be normally distributed with zero mean and variance σ^2.

4.3 Analysis using Stata

Assuming the data are in an ASCII file *bp.raw*, exactly as shown in Table 4.1, i.e., 12 rows, the first containing the observations 170 186 180 173 189 202, they can be read into Stata by producing a dictionary file, *bp.dct* containing the following statements:

```
dictionary using bp.raw{
   _column(6) int bp11
   _column(14) int bp12
   _column(22) int bp13
   _column(30) int bp01
   _column(38) int bp02
   _column(46) int bp03
}
```

and using the following command

```
infile using bp
```

Note that it was not necessary to define a dictionary here since the same result could have been achieved using a simple **infile** command (see exercises). The final dataset should have a single variable, **bp**, that contains all the blood pressures, and three additional variables, **drug**, **biofeed**, and **diet** representing the corresponding levels of drug, biofeedback and diet.

First, create **diet** which should take on one value for the first six rows and another for the following rows. This is achieved using the commands

```
gen diet=0 if _n<=6
replace diet=1 if _n>6
```

·or, more concisely, using

```
gen diet=_n>6
```

Now use the **reshape long** command to stack the columns on top of each other. If we specify bp0 and bp1 as the variable names in the **reshape** command, then bp01, bp02, and bp03 are stacked into one column with variable name bp0 (and similarly for bp1) and another variable is created that contains the suffixes 1, 2, and 3. We ask for this latter variable to be called **drug** using the option j(drug) as follows:

```
gen id=_n
reshape long bp0 bp1, i(id) j(drug)
list in 1/9
```

	id	drug	diet	bp0	bp1
1.	1	1	0	173	170
2.	1	2	0	189	186
3.	1	3	0	202	180
4.	2	1	0	194	175
5.	2	2	0	194	194
6.	2	3	0	228	187
7.	3	1	0	197	165
8.	3	2	0	217	201
9.	3	3	0	190	199

Here, **id** was generated because we needed to specify the row indicator in the i() option.

We now need to run the **reshape long** command again to stack up the columns bp0 and bp1 and generate the variable **biofeed**. The instructions to achieve this and to label all the variables are given below.

```
replace id=_n
reshape long bp, i(id) j(biofeed)
replace id=_n

label drop _all
label define d 0 "absent" 1 "present"
label values diet d
label values biofeed d
label define dr 1 "Drug X" 2 "Drug Y" 3 "Drug Z"
label values drug dr
```

To begin, it will be helpful to look at some summary statistics for each of the cells of the design. A simple way of obtaining the required summary measures is to use the **table** instruction.

```
table drug, contents(freq mean bp median bp sd bp)/*
  */ by(diet biofeed)
```

```
----------+-------------------------------------------------
diet,     |
biofeed   |
and drug  |      Freq.     mean(bp)      med(bp)        sd(bp)
----------+-------------------------------------------------
absent    |
absent    |
   Drug X |        6           188           192      10.86278
   Drug Y |        6           200           197      10.07968
   Drug Z |        6           209           205       14.3527
----------+-------------------------------------------------
absent    |
present   |
   Drug X |        6           168         167.5      8.602325
   Drug Y |        6           204           205      12.68069
   Drug Z |        6           189         190.5      12.61745
----------+-------------------------------------------------
present   |
absent    |
   Drug X |        6           173           172      9.797959
   Drug Y |        6           187           188      14.01428
   Drug Z |        6           182           179       17.1114
----------+-------------------------------------------------
present   |
present   |
   Drug X |        6           169           167      14.81891
   Drug Y |        6           172           170      10.93618
   Drug Z |        6           173         176.5       11.6619
----------+-------------------------------------------------
```

Display 4.1

The standard deviations in Display 4.1 indicate that there are considerable differences in the within cell variability. This may have implications for the analysis of variance of these data since one of the assumptions made is that the observations within each cell have the same variance. To begin, however, apply the model specified in Section 3.2 to the raw data using the **anova** instruction

```
anova bp drug diet biofeed diet*drug diet*biofeed /*
  */ drug*biofeed drug*diet*biofeed
```

The resulting ANOVA table is shown in Display 4.2.

The **Root MSE** is simply the square root of the residual mean square, with **R-squared** and **Adj R-squared** being as described in Chapter 3. The F-statistic of each effect represents the mean sum of squares for that effect, divided by the residual mean sum of squares, given under the heading **MS**. The main effects of drug ($F_{2,60} = 11.73$, $p < 0.001$), diet ($F_{1,60} = 33.20$, $p < 0.001$), and biofeed

```
                    Number of obs =      72    R-squared     =  0.5840
                    Root MSE      = 12.5167    Adj R-squared =  0.5077

         Source |  Partial SS    df      MS             F      Prob > F
    ------------+------------------------------------------------------
          Model |  13194.00      11  1199.45455       7.66     0.0000
                |
           drug |   3675.00       2   1837.50        11.73     0.0001
           diet |   5202.00       1   5202.00        33.20     0.0000
        biofeed |   2048.00       1   2048.00        13.07     0.0006
      diet*drug |    903.00       2    451.50         2.88     0.0638
   diet*biofeed |     32.00       1     32.00         0.20     0.6529
   drug*biofeed |    259.00       2    129.50         0.83     0.4425
drug*diet*biofeed|  1075.00       2    537.50         3.43     0.0388
                |
       Residual |   9400.00      60   156.666667
    ------------+------------------------------------------------------
          Total |  22594.00      71   318.225352
```

Display 4.2

($F_{1,60} = 13.07$, $p < 0.001$) are all highly significant, and the three way interaction drug*diet*biofeed is also significant at the 5% level. The existence of a three-way interaction complicates the interpretation of the other terms in the model; it implies that the interaction between any two of the factors is different at the different levels of the third factor. Perhaps the best way of trying to understand the meaning of the three-way interaction is to plot a number of *interaction diagrams*; that is, plots of mean values for a factor at the different levels of the other factors.

This can be done by first creating a variable predbp containing the predicted means (which in this case coincide with the observed cell means because the model is saturated) using the command

```
predict predbp
```

In order to produce a graph with separate lines for the factor diet, we need to generate the variables bp0 and bp1 as follows:

```
gen bp0=predbp if diet==0
label variable bp0 "diet absent"
gen bp1=predbp if diet==1
label variable bp1 "diet present"
```

Plots of predbp against biofeed for each level of drug with seperate lines for diet can be obtained using the instructions

```
sort drug
graph bp0 bp1 biofeed, sort c(ll) by(drug) xlab(0,1) ylab /*
     */ b2(" ") b1("Blood pressure against biofeedback")
```

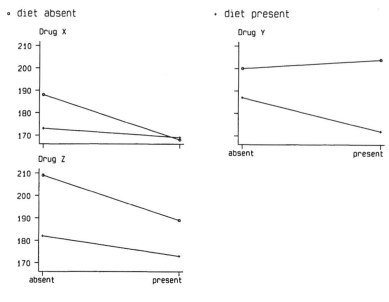

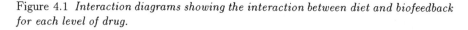

Blood pressure against biofeedback

Figure 4.1 *Interaction diagrams showing the interaction between diet and biofeedback for each level of drug.*

The resulting interaction diagrams are shown in Figure 4.1. For drug Y, the presence of biofeedback increases the effect of diet; whereas for drug Z the effect of diet is hardly altered by the presence of biofeedback and for drug X the effect is decreased.

Tables of the cell means plotted in the interaction diagrams, as well as the corresponding standard deviations, are produced for each drug using the following command:

```
table diet biofeed, contents(mean bp sd bp) by(drug)
```

giving the outpit shown in Display 4.3.

As mentioned previously, the observations in the 12 cells of the $3 \times 2 \times 2$ design have variances which differ considerably. Consequently, an analysis of variance of the data transformed in some way might be worth considering. For example, to analyze the log transformed observations, we can use the following instructions:

```
gen lbp=log(bp)
anova lbp drug diet biofeed diet*drug diet*biofeed /*
    */ drug*biofeed drug*diet*biofeed
```

```
----------+--------------------
drug and  |      biofeed
diet      |   absent    present
----------+--------------------
Drug X    |
  absent  |      188        168
          | 10.86278   8.602325
          |
  present |      173        169
          | 9.797959  14.81891
----------+--------------------
Drug Y    |
  absent  |      200        204
          | 10.07968  12.68069
          |
  present |      187        172
          | 14.01428  10.93618
----------+--------------------
Drug Z    |
  absent  |      209        189
          | 14.3527   12.61745
          |
  present |      182        173
          | 17.1114   11.6619
----------+--------------------
```

Display 4.3

```
                    Number of obs =      72    R-squared       =  0.5776
                    Root MSE      = .068013    Adj R-squared   =  0.5002

          Source |  Partial SS     df       MS            F      Prob > F
    -------------+----------------------------------------------------------
           Model |  .379534762     11    .03450316        7.46    0.0000
                 |
            diet |  .149561559      1   .149561559       32.33    0.0000
            drug |  .107061236      2   .053530618       11.57    0.0001
         biofeed |  .061475507      1   .061475507       13.29    0.0006
       diet*drug |  .024011594      2   .012005797        2.60    0.0830
    diet*biofeed |  .000657678      1   .000657678        0.14    0.7075
    drug*biofeed |  .006467873      2   .003233936        0.70    0.5010
diet*drug*biofeed|  .030299315      2   .015149657        3.28    0.0447
                 |
        Residual |  .277545987     60   .004625766
    -------------+----------------------------------------------------------
           Total |  .657080749     71   .009254658
```

The result is similar to the untransformed blood pressures. Since the three way interaction is only marginally significant and if no substantive explanation of this interaction is available, it might be better to interpret the results in terms of the very significant main effects. The relevant summary statistics for the log transformed blood pressures can be obtained using the following instructions:

```
table drug, contents(mean lbp sd lbp)
```

```
----------+--------------------------
    drug |   mean(lbp)        sd(lbp)
----------+--------------------------
  Drug X |    5.159152         .075955
  Drug Y |    5.247087        .0903675
  Drug Z |    5.232984        .0998921
----------+--------------------------
```

```
table diet, contents(mean lbp sd lbp)
```

```
----------+--------------------------
    diet |   mean(lbp)        sd(lbp)
----------+--------------------------
  absent |    5.258651        .0915982
 present |    5.167498        .0781686
----------+--------------------------
```

```
table biofeed, contents(mean lbp sd lbp)
```

```
----------+--------------------------
 biofeed |   mean(lbp)        sd(lbp)
----------+--------------------------
  absent |    5.242295        .0890136
 present |    5.183854        .0953618
----------+--------------------------
```

Drug X appears to produce lower blood pressures as does the special diet and the presence of biofeedback. Readers are encouraged to try other transformations.

Note that the model with main effects only can also be estimated using regression with dummy variables. Since drug has three levels and therefore requires two dummy variables, we save some time by using the xi: prefix as follows:

```
xi: regress lbp i.drug i.diet i.biofeed
```

```
i.drug              Idrug_1-3     (naturally coded; Idrug_1 omitted)
i.diet              Idiet_0-1     (naturally coded; Idiet_0 omitted)
i.biofeed           Ibiofe_0-1    (naturally coded; Ibiofe_0 omitted)
```

```
  Source |      SS         df       MS              Number of obs =      72
---------+------------------------------            F(  4,    67) =   15.72
   Model | .318098302      4    .079524576          Prob > F      =  0.0000
Residual | .338982447     67    .00505944           R-squared     =  0.4841
---------+------------------------------            Adj R-squared =  0.4533
   Total | .657080749     71    .009254658          Root MSE      =  .07113
```

```
     lbp |    Coef.    Std. Err.       t     P>|t|     [95% Conf. Interval]
---------+--------------------------------------------------------------------
  Idrug_2 |  .0879354   .0205334     4.283   0.000     .0469506    .1289203
  Idrug_3 |  .0738315   .0205334     3.596   0.001     .0328467    .1148163
  Idiet_1 | -.0911536   .0167654    -5.437   0.000    -.1246175   -.0576896
 Ibiofe_1 | -.0584406   .0167654    -3.486   0.001    -.0919046   -.0249767
    _cons |  5.233949   .0187443   279.228   0.000     5.196535    5.271363
```

The coefficients represent the mean differences between each level compared with the reference level (the omitted categories: drug X, diet absent, and biofeedback absent) when the other variables are held constant. Due to the balanced nature of the design (see next chapter), the p-values are equal to those of ANOVA except that no overall p-value for drug is given. This can be obtained using

```
testparm Idrug*
```

```
( 1)  Idrug_2 = 0.0
( 2)  Idrug_3 = 0.0

      F(  2,    67) =    10.58
           Prob > F =     0.0001
```

The F-statistic is different from that in the last anova command because no interactions were included in the model so that the residual degrees of freedom and the residual sum of squares have both increased.

4.4 Exercises

1. Reproduce the result of the command infile using bp without using the dictionary and follow the reshape instructions to generate the required dataset.

2. Produce three diagram with boxplots (1) for each level of drug, (2) for each level of diet, and (3) for each level of biofeedback.

3. Investigate other possible transformations of the data.

4. Suppose that in addition to the blood pressure of each of the individuals in the study, the investigator had also recorded their ages in file *age.dat* with

Table 4.2: Ages in age.dat to be used as a covariate

id	age	id	age
1	39	37	45
2	39	38	58
3	61	39	61
4	50	40	47
5	51	41	67
6	43	42	49
7	59	43	54
8	50	44	48
9	47	45	46
10	60	46	67
11	77	47	56
12	57	48	54
13	62	49	66
14	44	50	43
15	63	51	47
16	77	52	35
17	56	53	50
18	62	54	60
19	44	55	73
20	61	56	46
21	66	57	59
22	52	58	65
23	53	59	49
24	54	60	52
25	40	61	40
26	62	62	80
27	68	63	46
28	63	64	63
29	47	65	56
30	70	66	58
31	57	67	53
32	51	68	56
33	70	69	64
34	57	70	57
35	64	71	60
36	66	72	48

the results shown in Table 4.2. Reanalyze the data using age as a covariate
(see help merge and help anova).

CHAPTER 5

Analysis of Variance II: Effectiveness of Slimming Clinics

5.1 Description of data

Slimming clinics aim to help people lose weight by offering encouragement and support about dieting through regular meetings. In a study of their effectiveness, a 2 × 2 factorial design was used to investigate whether giving clients a technical manual containing slimming advice based on psychological behaviorist theory would help them to control their diet, and how this might be affected by whether or not a client had already been trying to slim. The data collected are shown in Table 5.1. (They are also given in Hand et al., 1994.) The response variable was defined as follows:

$$\frac{\text{weight at three months} - \text{ideal weight}}{\text{natural weight} - \text{ideal weight}} \tag{5.1}$$

Table 5.1: Data in slim.dat

cond	status	resp	cond	status	resp
1	1	-14.67	1	1	-1.85
1	1	-8.55	1	1	-23.03
1	1	11.61	1	2	0.81
1	2	2.38	1	2	2.74
1	2	3.36	1	2	2.10
1	2	-0.83	1	2	-3.05
1	2	-5.98	1	2	-3.64
1	2	-7.38	1	2	-3.60
1	2	-0.94	2	1	-3.39
2	1	-4.00	2	1	-2.31
2	1	-3.60	2	1	-7.69
2	1	-13.92	2	1	-7.64
2	1	-7.59	2	1	-1.62
2	1	-12.21	2	1	-8.85
2	2	5.84	2	2	1.71
2	2	-4.10	2	2	-5.19
2	2	0.00	2	2	-2.80

The number of observations in each cell of the design is not the same, so this is an example of a *unbalanced* 2×2 design.

5.2 Analysis of variance model

A suitable analysis of variance model for the data is

$$y_{ijk} = \mu + \alpha_i + \beta_j + \gamma_{ij} + \epsilon_{ijk} \qquad (5.2)$$

where y_{ijk} represents the weight change of the kth individual having status j and condition i, μ is the overall mean, α_i represents the effect of condition i, β_j the effect of status j, γ_{ij} the status × condition interaction, and ϵ_{ijk} the residuals–these are assumed to have a normal distribution with variance σ^2.

The unbalanced nature of the slimming data presents some difficulties for analysis not encountered in factorial designs having the same number of observations in each cell (see the previous chapter). The main problem is that when the data are unbalanced there is no unique way of finding a 'sum of squares' corresponding to each main effect and their interactions, since these effects are no longer independent of one another. If the data were balanced, the among cells sum of squares would partition orthogonally into three component sums of squares representing the two main effects and their interaction. Several methods have been proposed for dealing with this problem and each leads to a different partition of the overall sum of squares. The different methods for arriving at the sums of squares for unbalanced designs can be explained in terms of the comparisons of different sets of specific models. For a design with two factors A and B, Stata can calculate the following types of sums of squares.

5.2.1 Sequential sums of squares

Sequential sums of squares (also known as hierarchical) represent the effect of adding a term to an existing model. So, for example, a set of sequential sums of squares such as

Source	SS
A	SSA
B	SSB|A
AB	SSAB|A,B

represent a comparison of the following models:

- SSAB|A,B—model including an interaction and main effects compared with one including only main effects.

- SSB|A—model including both main effects, but with no interaction, compared with one including only the main effects of factor A.

- SSA—model containing only the A main effect compared with one containing only the overall mean.

The use of these sums of squares in a series of tables in which the effects are considered in different orders (see later) will often provide the most satisfactory way of deciding which model is most appropriate for the observations. (These are SAS Type I sums of squares—see Everitt and Der, 1996.)

5.2.2 Unique sums of squares

By default, Stata produces unique sums of squares that represent the contribution of each term to a model including all the other terms. So, for a two-factor design, the sums of squares represent the following.

Source	SS
A	SSA|B,AB
B	SSB|A,AB
AB	SSAB|A,B

(These are SAS Type III sums of squares.) Note that these sums of squares generally do not add up to the total sums of squares.

5.2.3 Regression

As we have shown in Chapter 4, ANOVA models may also be estimated using regression by defining suitable dummy variables. Assume that A is represented by a single dummy variable. The regression coefficients for A represents the *partial* contribution of that variable, adjusted for all other variables in the model, say B. This is equivalent to the contribution of A to a model already including B. A complication with regression models is that, in the presence of an interaction, the p-values of the terms depend on the exact coding of the dummy variables (see Aitkin, 1978). The unique sums of squares correspond to regression where dummy variables are coded in a particular way, for example a two-level factor can be coded as -1,1.

There have been numerous discussions over which sums of squares are most appropriate for the analysis of unbalanced designs. The Stata manual appears to recommend its default for general use. Nelder (1977) and Aitkin (1978) however, are strongly critical of 'correcting' main effects for an interaction term involving the same factor; their criticisms are based on both theoretical and pragmatic arguments and seem compelling. A frequently used approach is therefore to test the highest order interaction adjusting for all lower order interactions and not vice versa. Both Nelder and Aitkin prefer the use of Type I sums of squares in association with different orders of effects as the procedure most likely to identify an appropriate model for a data set. For a detailed explanation of the various types of sums of squares, see Boniface (1995).

5.3 Analysis using Stata

The data can be read in from an ASCII file, *slim.dat* in the usual way using

```
infile cond status resp using slim.dat
```

A table showing the unbalanced nature of the 2×2 design can be obtained from

```
tabulate cond status
```

```
          | status
     cond |        1         2 |     Total
----------+--------------------+----------
        1 |        5        12 |        17
        2 |       11         6 |        17
----------+--------------------+----------
    Total |       16        18 |        34
```

We now use the **anova** command with no options specified to obtain the unique (Type III) sums of squares:

```
anova resp cond status cond*status
```

```
                    Number of obs =      34      R-squared     =  0.2103
                    Root MSE      = 5.9968      Adj R-squared =  0.1313

        Source |  Partial SS    df       MS            F      Prob > F
    -----------+----------------------------------------------------------
         Model |  287.231861     3  95.7439537         2.66    0.0659
               |
          cond |  2.19850409     1  2.19850409         0.06    0.8064
        status |  265.871053     1  265.871053         7.39    0.0108
   cond*status |  .130318264     1  .130318264         0.00    0.9524
               |
      Residual |  1078.84812    30   35.961604
    -----------+----------------------------------------------------------
         Total |  1366.07998    33  41.3963631
```

Our recommendation is that the sums of squares shown in this table are *not* used to draw inferences because the main effects have been adjusted for the interaction.

Instead we prefer an analysis that consists of obtaining two sets of sequential sums of squares, the first using the order **cond status cond*status** and the second the order **status cond cond*status**; the necessary instructions are

```
anova resp cond status cond*status, sequential
```

```
                    Number of obs =       34      R-squared     =  0.2103
                    Root MSE      = 5.9968      Adj R-squared =  0.1313

    Source |    Seq. SS     df       MS            F      Prob > F
-----------+---------------------------------------------------------
     Model | 287.231861      3  95.7439537        2.66     0.0659
           |
      cond |  21.1878098     1  21.1878098        0.59     0.4487
    status | 265.913733      1  265.913733        7.39     0.0108
cond*status|  .130318264     1  .130318264        0.00     0.9524
           |
  Residual | 1078.84812     30  35.961604
-----------+---------------------------------------------------------
     Total | 1366.07998     33  41.3963631
```

anova resp status cond cond*status, sequential

```
                    Number of obs =       34      R-squared     =  0.2103
                    Root MSE      = 5.9968      Adj R-squared =  0.1313

    Source |    Seq. SS     df       MS            F      Prob > F
-----------+---------------------------------------------------------
     Model | 287.231861      3  95.7439537        2.66     0.0659
           |
    status | 284.971071      1  284.971071        7.92     0.0085
      cond |  2.13047169     1  2.13047169        0.06     0.8094
cond*status|  .130318264     1  .130318264        0.00     0.9524
           |
  Residual | 1078.84812     30  35.961604
-----------+---------------------------------------------------------
     Total | 1366.07998     33  41.3963631
```

The sums of squares corresponding to model and residuals are, of course, the same in both tables, as is the sum of squares for the interaction term. What differs are the sums of squares in the cond and status rows in the two tables. The terms of most interest are the sum of squares of status|cond which is obtained from the table as 265.91, and the sum of squares of cond|status which is 2.13. These sums of squares are less than the sums of squares for status and cond alone (284.97 and 21.19, respectively), by an amount of 19.06, a portion of the sums of squares which cannot be uniquely attributed to either of the variables. The associated F-tests in the two tables make it clear that there is no interaction effect and that status|cond is significant but cond|status is not. The conclusion is that only status, i.e., whether or not the woman had been slimming for over one year, is important in determining weight change. Provision of the manual appears to have no discernible effect.

Results equivalent to the unique (Type III) sums of squares can be obtained using regression:

```
gen cond1=cond
recode cond1 1=-1 2=1
```

```
gen status1=status
recode status1 1=-1 2=1
gen statcond = cond1*status1
regress res cond1 status1 statcond
```

Source	SS	df	MS				
				Number of obs =	34		
				F(3, 30) =	2.66		
Model	287.231861	3	95.7439537	Prob > F =	0.0659		
Residual	1078.84812	30	35.961604	R-squared =	0.2103		
				Adj R-squared =	0.1313		
Total	1366.07998	33	41.3963631	Root MSE =	5.9968		

resp	Coef.	Std. Err.	t	P>|t|	[95% Conf. Interval]	
cond1	.2726251	1.102609	0.247	0.806	-1.979204	2.524454
status1	2.998042	1.102609	2.719	0.011	.746213	5.24987
statcond	-.066375	1.102609	-0.060	0.952	-2.318204	2.185454
_cons	-3.960958	1.102609	-3.592	0.001	-6.212787	-1.70913

These results differ from the regression used by Stata's **anova** with the option **regress**:

```
anova resp cond status cond*status, regress
```

Source	SS	df	MS				
				Number of obs =	34		
				F(3, 30) =	2.66		
Model	287.231861	3	95.7439537	Prob > F =	0.0659		
Residual	1078.84812	30	35.961604	R-squared =	0.2103		
				Adj R-squared =	0.1313		
Total	1366.07998	33	41.3963631	Root MSE =	5.9968		

resp	Coef.	Std. Err.	t	P>|t|	[95% Conf. Interval]	
_cons	-.7566666	2.448183	-0.309	0.759	-5.756524	4.24319
cond						
1	-.4125001	2.9984	-0.138	0.891	-6.536049	5.711049
2	(dropped)					
status						
1	-5.863333	3.043491	-1.927	0.064	-12.07897	.3523044
2	(dropped)					
cond*status						
1 1	-.2655002	4.410437	-0.060	0.952	-9.272815	8.741815
1 2	(dropped)					
2 1	(dropped)					
2 2	(dropped)					

because this uses different dummy variables. The dummy variables are equal to 1 for the levels to the left of the reported coefficient and zero otherwise, i.e., the dummy variable for **cond*status** is one when **status** and **cond** are both 1.

A table of mean values helpful in interpreting these results can be found using

```
table cond status, c(mean resp) row col f(%8.2f)
```

```
----------+--------------------
          |        status
     cond |    1        2  Total
----------+--------------------
        1 | -7.30    -1.17  -2.97
        2 | -6.62    -0.76  -4.55
          |
    Total | -6.83    -1.03  -3.76
----------+--------------------
```

The means demonstrate that experienced slimmers achieve the greatest weight reduction.

5.4 Exercises

1. Investigate what happens to the sequential sums of squares if the cond*status interaction term is given before the main effects cond status in the anova command with the sequential option.

2. Use regress to reproduce the analysis of variance by coding both condition and status as (0,1) dummy variables and creating an interaction variable as the product of these dummy variables.

3. Use regress in conjunction with xi: to fit the same model without the need to generate any dummy variables (see help xi:).

4. Reproduce the results of anova resp cond status cond*status, regress using regress by making xi: omit the last category instead of the first (see help xi).

See also the Exercises in Chapters 7 and 13.

CHAPTER 6

Logistic Regression: Treatment of Lung Cancer and Diagnosis of Heart Attacks

6.1 Description of data

Two datasets will be analyzed in this chapter. The first dataset shown in Table 6.1 originates from a clinical trial in which lung cancer patients were randomized to receive two different kinds of chemotherapy (sequential therapy and alternating therapy). The outcome was classified into one of four categories: progressive disease, no change, partial remission, or complete remission. The data were published in Holtbrugge and Schumacher (1991) and also appear in Hand et al. (1994). The main aim of any analysis will be to assess differences between the two therapies.

Table 6.1: Tumor data

Therapy	Sex	Progressive disease	No change	Partial remission	Complete remission
Sequential	Male	28	45	29	26
	Female	4	12	5	2
Alternative	Male	41	44	20	20
	Female	12	7	3	1

The second dataset in this chapter arises from a study investigating the use of Serum Creatine Kinase (CK) levels for the diagnosis of myocardial infarction (heart attack). Patients admitted to a coronary care unit because they were suspected of having had a myocardial infarction within the last 48 hours had their CK levels measured on admission and the next two mornings. A clinician who was "blind" to the CK results came to an independent "gold standard" diagnosis using electrocardiograms, clinical records, and autopsy reports. The maximum CK levels for 360 patients are given in Table 6.2 together with the clinician's diagnosis. The table was taken from Sackett et al. (1991), where only the ranges of CK levels were given, not their precise values.

The main questions of interest here is how well CK discriminates between those with and without myocardial infarction, and to investigate the characteristics of the diagnostic test for different thresholds.

Table 6.2: Diagnosis of myocardial infarction from Serum Creatine Kinase (CK)(Taken from Sackett et al.(1991) with permission of the publisher, Little Brown & Company.)

Maximum CK level	Infarct present	Infarct absent
0-39	2	88
40-79	13	26
80-119	30	8
120-159	30	5
160-199	21	0
200-239	19	1
240-279	18	1
280-319	13	1
320-359	19	0
360-399	15	0
400-439	7	0
440-479	8	0
480-	35	0

6.2 The logistic regression model

Logistic regression is used when the response variable is dichotomous. In this case, we are interested in how the probability that the response variable takes on the value of interest (usually coded as 1, the other value being zero) depends on a number of explanatory variables. The probability π that $y = 1$ is just the expected value of y. In linear regression (see Chapter 2), the expected value of y is modeled as a linear function of the explanatory variables,

$$E[y] = \beta_0 + \beta_1 x_1 + \beta_2 x_2 + \cdots + \beta_p x_p. \tag{6.1}$$

However, there are two problems with using the method of linear regression when y is dichotomous: (1) the predicted probability must satisfy $0 \leq \pi \leq 1$ whereas a linear predictor can yield any value from minus infinity to plus infinity, and (2) the observed values of y do not follow a normal distribution with mean π, but rather a Bernoulli (or Binomial$(1,\pi)$) distribution.

In logistic regression, the first problem is addressed by replacing the probability $\pi = E[y]$ on the left-hand side of equation (6.1) by the *logit* of the probability, giving

$$\text{logit}(\pi) = \log(\pi/(1 - \pi)) = \beta_0 + \beta_1 x_1 + \beta_2 x_2 + \cdots + \beta_p x_p. \tag{6.2}$$

The logit of the probability is simply the log of the odds of the event of interest. Writing β and $\mathbf{x}_i$ for the column vectors $(\beta_0, \cdots, \beta_p)^T$ and $(1, x_{1i}, \cdots, x_{pi})^T$, respectively, the predicted probability as a function of the linear predictor is

$$\pi(\beta^T \mathbf{x}_i) = \frac{\exp(\beta^T \mathbf{x}_i)}{1 + \exp(\beta^T \mathbf{x}_i)}$$

$$= \frac{1}{1 + \exp(-\beta^T \mathbf{x}_i)}. \tag{6.3}$$

When the logit takes on any real value, this probability always satisfies $0 \leq \pi(\beta^T \mathbf{x}_i) \leq 1$.

The second problem relates to the estimation procedure. Whereas maximum likelihood estimation in linear regression leads to (weighted) least squares, this is not the case in logistic regression. The log likelihood function for logistic regression is

$$l(\beta; \mathbf{y}) = \sum_i [y_i \log(\pi(\beta^T \mathbf{x}_i)) + (1 - y_i) \log(1 - \pi(\beta^T \mathbf{x}_i))]. \tag{6.4}$$

This log likelihood is maximized numerically using an iterative algorithm. For full details of logistic regression, see for example Collett (1991).

Logistic regression can be generalized to the situation where the response variable has more than two ordered response categories $y_1, \cdots, y_I$ by thinking of these categories as resulting from thresholding an unobserved continuous variable $S = \beta^T \mathbf{x} + u$ at a number of cut-points κ_j, $j = 1, \cdots, I - 1$ so that $y = y_1$ if $S \leq \kappa_1$, $y = y_2$ if $\kappa_1 < S \leq \kappa_2$, ... and $y = y_I$ if $\kappa_{I-1} < S$. The variable u is assumed to have the standard logistic distribution $\Pr(u \leq X) = 1/(1+\exp(-X))$ so that the cumulative probability γ_j of a response up to and including y_j, $\Pr(y \leq y_j)$, is

$$\gamma_j = \Pr(S \leq \kappa_j) = \Pr(S - \beta^T \mathbf{x} \leq \kappa_j - \beta^T \mathbf{x})$$

$$= \frac{1}{1 + \exp(\beta^T \mathbf{x} - \kappa_j)} \tag{6.5}$$

where κ_I is taken to be ∞ so that $\gamma_I = 1$. The probability of the jth response category is then simply $\pi_j = \gamma_j - \gamma_{j-1}$.

The ordinal regression model is called the *proportional odds model* because the log odds that $y > y_j$ is

$$\log(\frac{1 - \gamma_j(x)}{\gamma_j(x)}) = \beta^T \mathbf{x} - \kappa_j \tag{6.6}$$

so that the log odds ratio for two values of $\mathbf{x}$ is $\beta^T (\mathbf{x}_1 - \mathbf{x}_2)$ and is independent of j. Therefore, $\exp(\beta_k)$ is the odds ratio that $y > y_j$ for any j when x_k increases by 1.

In ordinary logistic regression, $y_1 = 0$, $y_2 = 1$, $\kappa_1 = 0$ and $\exp(\beta_k)$ is the odds ratio that $y = 1$ when x_k increases by 1.

Note that the probit and ordinal probit models correspond to logistic and

ordinal logistic regression models with the cumulative distribution function
in (6.5) replaced by the standard normal distribution.

6.3 Analysis using Stata

6.3.1 Chemotherapy treatment of lung cancer

Assume the ASCII file *tumor.dat* contains the four by four matrix of frequencies
shown in Table 6.1. First read the data and generate variables for therapy and
sex:

```
infile fr1 fr2 fr3 fr4 using tumor.dat
gen therapy=int((_n-1)/2)
sort therapy
by therapy: gen sex=_n
label define t 0 seq 1 alt
label values therapy t
label define s 1 male 2 female
label values sex s
```

Then reshape the data to long, placing the four levels of the outcome into a
variable outc, and expand the dataset to have one observation per subject:

```
reshape long fr, i(therapy sex) j(outc)
expand fr
```

We can check that the data conversion is correct by tabulating these data
as in Table 6.1

```
table sex outc, contents(freq) by(therapy)
```

```
----------+-----------------------------
therapy   |           outc
and sex   |    1     2     3     4
----------+-----------------------------
seq       |
     male |   28    45    29    26
   female |    4    12     5     2
----------+-----------------------------
alt       |
     male |   41    44    20    20
   female |   12     7     3     1
----------+-----------------------------
```

In order to be able to carry out ordinary logistic regression, we need to
dichotomize the outcome, for example, by considering partial and complete
remission to be an improvement and the other categories to be no improvement.
The new outcome variable may be generated as follows:

```
gen improve=outc
recode improve 1/2=0 3/4=1
```

or using

```
gen improve = outc>2
```

The command logit for logistic regression follows the same syntax as regress and all other estimation commands. For example, automatic selection procedures can be carried out using sw and post-estimation commands such as testparm are available. First, include therapy as the only explanatory variable:

```
logit improve therapy
```

```
Iteration 0:   Log Likelihood =-194.40888
Iteration 1:   Log Likelihood =-192.30753
Iteration 2:   Log Likelihood =-192.30471

Logit Estimates                                  Number of obs =      299
                                                 chi2(1)       =     4.21
                                                 Prob > chi2   = 0.0402
Log Likelihood = -192.30471                      Pseudo R2     = 0.0108

------------------------------------------------------------------------------
improve |    Coef.   Std. Err.       z     P>|z|     [95% Conf. Interval]
---------+--------------------------------------------------------------------
therapy |  -.4986993  .2443508    -2.041   0.041     -.977618   -.0197805
  _cons |  -.361502   .1654236    -2.185   0.029    -.6857263   -.0372777
------------------------------------------------------------------------------
```

The algorithm takes three iterations to converge. The coefficient of therapy represents the difference in the log odds (of an improvement) between the alternating and sequential therapies. The negative value indicates that sequential therapy is superior to alternating therapy. The p-value of the coefficient is 0.041 in the table. This was derived from the Wald-statistic, z, which is equal to the coefficient divided by its asymptotic standard error (Std. Err.) as derived from the Hessian matrix of the log likelihood function, evaluated at the maximum likelihood solution. This p-value is less reliable than the p-value based on the likelihood ratio between the model including only the constant and the current model, which is given at the top of the output (chi2(1)=4.21). Here, minus twice the log of the likelihood ratio is equal to 4.21 which has an approximate χ^2-distribution with one degree of freedom (because there is one additional parameter) giving a p-value of 0.040, very similar to that based on the Wald test. The coefficient of therapy represents the difference in log odds between the therapies and is not easy to interpret apart from the sign. Taking the exponential of the coefficient gives the odds ratio and exponentiating the 95% confidence limits gives the confidence interval for the odds ratio. Fortunately, the command logistic may be used to obtain the required odds ratio and its confidence interval

```
logistic improve therapy
```

```
Logit Estimates                              Number of obs =      299
                                             chi2(1)        =     4.21
                                             Prob > chi2    = 0.0402
Log Likelihood = -192.30471                  Pseudo R2      = 0.0108

-------------------------------------------------------------------------
improve | Odds Ratio  Std. Err.       z    P>|z|     [95% Conf. Interval]
--------+----------------------------------------------------------------
therapy |  .6073201   .1483991    -2.041   0.041     .3762061    .9804138
-------------------------------------------------------------------------
```

The standard error now represents the approximate standard error of the odds
ratio (calculated using the delta method). However, the Wald statistic and
confidence interval are derived using the log odds and its standard error, as
the sampling distribution of the log odds is much closer to normal than that
of the odds. To be able to test whether the inclusion of **sex** in the model
significantly increases the likelihood, the current likelihood can be saved using

```
lrtest, saving(1)
```

Including sex gives

```
logistic improve therapy sex
```

```
Logit Estimates                              Number of obs =      299
                                             chi2(2)        =     7.55
                                             Prob > chi2    = 0.0229
Log Likelihood = -190.63171                  Pseudo R2      = 0.0194

-------------------------------------------------------------------------
improve | Odds Ratio  Std. Err.       z    P>|z|     [95% Conf. Interval]
--------+----------------------------------------------------------------
therapy |  .6051969   .1486907    -2.044   0.041     .3739084    .9795537
    sex |  .5197993   .1930918    -1.761   0.078     .2509785   1.076551
-------------------------------------------------------------------------
```

The p-value of **sex** based on the Wald-statistic is 0.078 and a p-value for the
likelihood ratio test is obtained using

```
lrtest, saving(0)
lrtest, model(1) using(0)
```

```
Logistic:  likelihood-ratio test          chi2(1)      =        3.35
                                          Prob > chi2 =      0.0674
```

which is not very different to the value of 0.078. Note that **model(1)** refers
to the model which is nested in the more complicated model referred to by
using(0). If the two models had been fitted in the reverse order, it would not
have been necessary to specify these options because **lrtest** assumes that the

current model is nested in the last model saved using `lrtest, saving(0)`. Retaining the variable **sex** in the model, the predicted probabilities can be obtained using

    predict prob

and the four different predicted probabilites may be compared with the observed proportions as follows:

    table sex , contents(mean prob mean improve freq) by(therapy)

```
----------+----------------------------------------------------
therapy   |
and sex   |      mean(prob)  mean(improve)              Freq.
----------+----------------------------------------------------
seq       |
    male  |       .4332747       .4296875                128
  female  |       .2843846       .3043478                 23
----------+----------------------------------------------------
alt       |
    male  |       .3163268            .32                125
  female  |       .1938763        .173913                 23
----------+----------------------------------------------------
```

The agreement is good, so there appears to be no strong interaction between sex and type of therapy. (We could test for an interaction between sex and therapy by using `xi: logistic improve i.therapy*i.sex`.) Residuals are not very informative for these data because there are only four different predicted probabilities.

We now fit the proportional odds model using the full ordinal response variable **outc**:

    ologit outc therapy sex, table

The results are shown in Display 6.1. Both **therapy** and **sex** are more significant than before. The coefficients represent the log odds ratios of being in complete remission versus being at best in partial remission; or equivalently, the log odds ratio of being at least in partial remission rather than having progressive disease or no change. The option **table** has produced the last part of the output to remind us how the proportional odds model is defined. We could calculate the probability that a male (**sex**=1) who is receiving sequential therapy (**therapy**=0) will be in complete remission (**outc**=4) using $1 - \gamma_3$ (see equation (6.5))

    display 1-1/(1+exp(-0.5413938-0.758662))

> .21415563

However, a much quicker way of computing the predicted probabilities for all four responses and all combinations of explanatory variables is to use the command **predict**

```
Iteration 0:  Log Likelihood =-399.98398
Iteration 1:  Log Likelihood =-394.53988
Iteration 2:  Log Likelihood =-394.52832

Ordered Logit Estimates                        Number of obs =    299
                                               chi2(2)       =  10.91
                                               Prob > chi2   = 0.0043
Log Likelihood = -394.52832                    Pseudo R2     = 0.0136

--------------------------------------------------------------------------
    outc |    Coef.   Std. Err.      z    P>|z|    [95% Conf. Interval]
---------+----------------------------------------------------------------
 therapy |  -.580685   .2121432   -2.737  0.006   -.9964781   -.164892
     sex |  -.5413938  .2871764   -1.885  0.059   -1.104249   .0214616
---------+----------------------------------------------------------------
   _cut1 |  -1.859437  .3828641         (Ancillary parameters)
   _cut2 |  -.2921603  .3672626
   _cut3 |   .758662   .3741486
--------------------------------------------------------------------------

    outc |      Probability          Observed
---------|-----------------------------------------
       1 |   Pr(       xb+u<_cut1)    0.2843
       2 |   Pr(_cut1<xb+u<_cut2)     0.3612
       3 |   Pr(_cut2<xb+u<_cut3)     0.1906
       4 |   Pr(_cut3<xb+u)           0.1639
```

Display 6.1

```
predict p1 p2 p3 p4
```

and to tabulate the results as follows:

```
table sex, contents(mean p1 mean p2 mean p3 mean p4) by(therapy)
```

```
----------+------------------------------------------------------
therapy   |
and sex   |   mean(p1)    mean(p2)    mean(p3)    mean(p4)
----------+------------------------------------------------------
seq       |
    male  |   .2111441    .3508438    .2238566    .2141556
  female  |   .3150425    .3729235    .175154     .1368799
----------+------------------------------------------------------
alt       |
    male  |   .3235821    .3727556    .1713585    .1323038
  female  |   .4511651    .346427     .1209076    .0815003
----------+------------------------------------------------------
```

6.3.2 Diagnosis of heart attacks

The data in *sck.dat* are read in using

```
infile ck pres abs using sck.dat
```

Each observation represents all subjects with maximum creatine kinase values in the same range. The total number of subjects is pres+abs

```
gen tot=pres+abs
```

and the number of subjects with the disease is pres. The probability associated with each observation is Binomial(tot, π). The programs logit and logistic are for data where each observation represents a single Bernouilli trial, with probability Binomial(1, π). Another program, blogit, can be used to analyze the "grouped" data with "denominators" tot:

```
blogit pres tot ck
```

```
Logit Estimates                              Number of obs =      360
                                             chi2(1)       =   283.15
                                             Prob > chi2   =   0.0000
Log Likelihood = -93.886407                  Pseudo R2     =   0.6013

---------------------------------------------------------------------
_outcome |    Coef.   Std. Err.      z     P>|z|    [95% Conf. Interval]
---------+-----------------------------------------------------------
      ck |  .0351044   .0040812    8.601   0.000    .0271053    .0431035
   _cons | -2.326272   .2993611   -7.771   0.000   -2.913009   -1.739535
---------------------------------------------------------------------
```

There is a very significant association between CK and the probability of infarct. We now need to investigate whether it is reasonable to assume that the log odds depends linearly on CK. Therefore, plot the observed proportions and predicted probabilities as follows:

```
gen prop = pres/tot
predict pred, p
label variable prop "observed"
label variable pred "predicted"
graph pred prop ck, c(s.) xlab ylab l1("Probability") gap(4)
```

The predict command gives predicted counts by default and therefore the p option was used to obtain predicted probabilities instead. The option c(s.) in the graph command causes the first set of points to be connected by a smooth curve. In the resulting graph in Figure 6.1, the curve fits the data reasonably well, the largest discrepancy being at CK=280.

We will now plot some residuals and then consider the sensitivity and specificity of CK as a diagnostic tool for different thresholds. There are some useful post-estimation commands available for these purposes for use after logistic that are not available after blogit. We therefore transform the data into the form required for logistic, i.e., one observation per Bernouilli trial with outcome infct equal to 0 or 1 so that the number of ones per CK level equals pres:

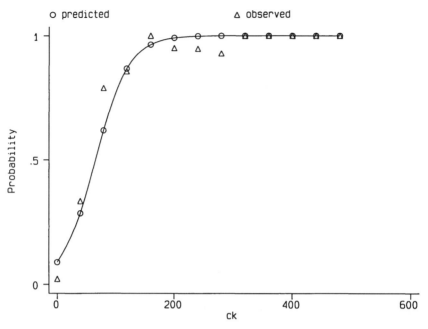

Figure 6.1 *Probability of infarct as a function of creatine Kinase levels.*

```
expand tot
sort ck
gen infct=0
quietly by ck: replace infct=1 if _n<=pres
```

We can reproduce the results of `blogit` using `logit`

```
logit infct ck, nolog
```

```
Logit Estimates                            Number of obs =      360
                                           chi2(1)       = 283.15
                                           Prob > chi2   = 0.0000
Log Likelihood = -93.886407                Pseudo R2     = 0.6013


------------------------------------------------------------------------------
  infct |      Coef.   Std. Err.       z     P>|z|      [95% Conf. Interval]
--------+---------------------------------------------------------------------
     ck |   .0351044   .0040812      8.601   0.000      .0271053     .0431035
  _cons |  -2.326272   .2993611     -7.771   0.000     -2.913009    -1.739535
------------------------------------------------------------------------------
```

where the `nolog` option was used to stop the iteration history being given.

One useful type of residual is the standardized Pearson residual for each "covariate pattern", i.e., for each combination of values in the covariates (here for each value of CK). These residuals may be obtained and plotted as follows:

```
predict resi, rstandard
graph resi ck, s([ck]) psize(150) xlab ylab gap(5)
```

The graph is shown in Figure 6.2. There are several large outliers. The largest

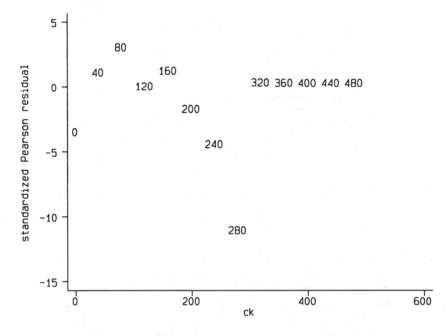

Figure 6.2 *Standardized Pearson residuals vs. creatine Kinase level.*

outlier at CK=280 is due to one subject out of 14 not having had an infarct although the predicted probability of an infarct is almost 1.

We now determine the accuracy of the diagnostic test. A *classification table* of the predicted diagnosis (using a cut-off of the predicted probability of 0.5) versus the true diagnosis may be obtained using

```
lstat
```

giving the table shown in Display 6.2. Both the sensitivity and specificity are relatively high. These characteristics are generally assumed to generalize to other populations whereas the positive and negative predictive values depend on the prevalence (or prior probability) of the condition (for example, see Sackett et al., 1991).

The use of other probability cut-offs could be investigated using the option cutoff(#) in the above command or using the commands lroc to plot a

```
Logistic model for infct

                -------- True --------
Classified |         D               ~D              Total
-----------+-------------------------------+-----------
    +      |       215              16  |      {    231
    -      |        15             114  |           129
-----------+-------------------------------+-----------
  Total    |       230             130  |           360

Classified + if predicted Pr(D) >= .5
True D defined as infct ~= 0
-----------------------------------------------------
Sensitivity                     Pr( +| D)    93.48%
Specificity                     Pr( -|~D)    87.69%
Positive predictive value       Pr( D| +)    93.07%
Negative predictive value       Pr(~D| -)    88.37%
-----------------------------------------------------
False + rate for true ~D        Pr( +|~D)    12.31%
False - rate for true D         Pr( -| D)     6.52%
False + rate for classified +   Pr(~D| +)     6.93%
False - rate for classified -   Pr( D| -)    11.63%
-----------------------------------------------------
Correctly classified                         91.39%
-----------------------------------------------------
```

Display 6.2

ROC-curve (specificity vs. sensitivity for different cut-offs) or **lsens** to plot sensitivity and specificity against cut-off (see exercises).

The above classification table may be misleading because we are testing the model on the same sample that was used to derive it. An alternative approach is to compute predicted probabilities for each observation from a model fitted to the remaining observations. This method, called "leave one out" method or *jackknifing* (see Lachenbruch and Mickey, 1986), can be carried out relatively easily for our data because we only have a small number of covariate and response patterns. Instead of looping through all observations, excluding each observation in the logistic regression command and computing that observation's predicted probability, we can loop through a subset of observations representing all combinations of covariates and responses found in the data.

First, label each unique covariate pattern consecutively in a variable **num** using **predict** with the **number** option:

```
predict num, number
```

Now generate **first**, equal to one for the first observation in each group of unique covariate and response patterns and zero otherwise:

```
sort num infct
quietly by num infct : gen first = _n==1
```

Now define **grp**, equal to the cumulative sum of **first**, obtained using the function **sum()**. This variable numbers the groups of unique covariate and response patterns consecutively:

```
gen grp=sum(first)
```

(An alternative way of generating **grp** without having to first create **first** would be to use the command **egen grp=group(num infct)**.) Now determine the number of unique combinations of CK levels and infarct status:

```
summ grp
```

```
Variable |     Obs      Mean   Std. Dev.       Min        Max
---------+-----------------------------------------------------
     grp |     360   8.658333   6.625051         1         20
```

Since there are 20 groups, we need to run **logistic** 20 times (for each value of **grp**), excluding one observation from **grp** to derive the model for predicting the probability for all observations in **grp**.

First generate a variable, **nxt**, that consecutively labels the 20 observations to be excluded in turn:

```
gen nxt = 0
replace nxt = grp if first==1
```

Now build up a variable **prp** of predicted probabilities as follows:

```
gen prp=0
for num 1/20:logistic infct ck if nxt~=X \ predict p \ /*
     */ replace prp=p if grp==X \ drop p
```

Here we used back-slashes to run four commands in each iteration of the loop, to

1. derive the model excluding one observation from **grp**
2. obtain the predicted probabilities p (**predict** produces results for the whole sample, not just the estimation sample)
3. set **prp** to the predicted probability for all observations in **grp** and
4. drop p so that it can be defined again in the next iteration.

The classification table for the jackknifed probabilities can be obtained using

```
gen class= prp>=0.5
tab class infct
```

```
          | infct
    class |         0          1 |     Total
----------+----------------------+----------
        0 |       114         15 |       129
        1 |        16        215 |       231
----------+----------------------+----------
    Total |       130        230 |       360
```

giving the same result as before (this is not generally the case).

6.4 Exercises

1. Read in the data without using the **expand** command and reproduce the result of ordinal logistic regressions by using the appropriate weights.

2. Test for an association between **depress** and **life** for the data described in Chapter 2 using

   (a) ordinal logistic regression with **depress** as dependent variable

   (b) logisitc regression with **life** as dependent variable.

3. Use **sw** together with **logistic** to find a model for predicting **life** using the data from Chapter 2 with different sets of candidate variables (see Chapter 3).

4. Produce a graph similar to that in Figure 6.1 using probit analysis (see help for **bprobit**).

5. Explore the use of **lstat**, **cutoff(#)**, **lroc**, and **lsens** for the diagnosis data.

CHAPTER 7

Generalized Linear Models: Australian School Children

7.1 Description of data

This chapter reanalyzes a number of datasets discussed in previous chapters and in addition, describes the analysis of a new dataset given in Aitkin (1978) . These data come from a sociological study of Australian Aboriginal and white children. The sample included children from four age groups (final year in primary school and first three years in secondary school) who were classified as slow or average learners. The number of days absent from school during the school year was recorded for each child. The data are given in Table 7.1. The variables are as follows:

- **eth**: ethnic group (A=aboriginal, N=white)
- **sex**: sex (M=male, F=female)
- **age**: class in school (F0, F1, F2, F3)
- **lrn**: average or slow learner(SL=slow learner, AL=average learner)
- **days**: number of days absent from school in one year

7.2 Generalized linear models

Previous chapters have described linear and logistic regression. In this chapter, we will describe an even more general class of models, called generalized linear models, of which linear regression and logistic regression are special cases.

Both linear and logistic regression involve a linear combination of the explanatory variables, called the linear predictor, of the form

$$
\begin{aligned}
\eta &= \beta_0 + \beta x_1 + \beta x_2 + \cdots + \beta x_p \\
&= \beta^T \mathbf{x}.
\end{aligned}
\tag{7.1}
$$

In both types of regression, the linear predictor determines the expectation μ of the response variable. In linear regression, where the response is continuous, μ is directly equated with the linear predictor. This is not advisable when the response is dichotomous because in this case the expectation is a probability which must satisfy $0 \leq \mu \leq 1$. In logistic regression, the linear predictor is

Table 7.1 Data in quine.dta presented in four columns to save space. (Taken from Aitkin (1978) with permission of the Royal Statistical Society.)

eth	sex	age	lrn	days	eth	sex	age	lrn	days	eth	sex	age	lrn	days	eth	sex	age	lrn	days
A	M	F0	SL	2	A	M	F0	SL	11	A	M	F0	SL	14	A	M	F0	AL	5
A	M	F0	AL	5	A	M	F0	AL	13	A	M	F0	AL	20	A	M	F0	AL	22
A	M	F1	SL	6	A	M	F1	SL	6	A	M	F1	SL	15	A	M	F1	AL	7
A	M	F1	AL	14	A	M	F2	SL	6	A	M	F2	SL	32	A	M	F2	SL	53
A	M	F2	SL	57	A	M	F2	AL	14	A	M	F2	AL	16	A	M	F2	AL	16
A	M	F2	AL	17	A	M	F2	AL	40	A	M	F2	AL	43	A	M	F2	AL	46
A	M	F3	SL	8	A	M	F3	AL	23	A	M	F3	AL	23	A	M	F3	AL	28
A	M	F3	AL	34	A	M	F3	AL	36	A	M	F3	AL	38	A	F	F0	AL	3
A	F	F0	AL	5	A	F	F0	AL	11	A	F	F0	AL	24	A	F	F0	AL	45
A	F	F1	SL	5	A	F	F1	SL	6	A	F	F1	SL	6	A	F	F1	SL	9
A	F	F1	SL	13	A	F	F1	SL	23	A	F	F1	SL	25	A	F	F1	SL	32
A	F	F1	AL	53	A	F	F1	AL	54	A	F	F1	AL	5	A	F	F1	AL	5
A	F	F2	SL	11	A	F	F2	SL	17	A	F	F2	SL	19	A	F	F2	SL	8
A	F	F2	AL	13	A	F	F2	AL	14	A	F	F2	AL	20	A	F	F2	SL	47
A	F	F3	SL	48	A	F	F3	SL	60	A	F	F3	AL	81	A	F	F3	AL	2
A	F	F3	AL	0	A	F	F3	AL	2	A	F	F3	AL	3	A	F	F3	AL	5
A	F	F3	AL	10	A	F	F3	AL	14	A	F	F3	SL	21	A	N	F3	SL	36
N	M	F0	AL	40	N	M	F0	SL	6	N	M	F0	AL	17	N	M	F0	AL	67
N	M	F0	SL	0	N	M	F0	SL	0	N	M	F0	SL	2	N	M	F0	SL	7
N	M	F1	SL	11	N	M	F1	SL	12	N	M	F1	SL	0	N	M	F1	SL	0
N	M	F1	SL	5	N	M	F1	AL	5	N	M	F1	SL	5	N	M	F2	SL	11
N	M	F1	AL	17	N	M	F1	AL	3	N	M	F2	AL	4	N	M	F2	AL	22
N	M	F2	SL	30	N	M	F2	SL	36	N	M	F2	AL	8	N	M	F2	AL	0
N	M	F2	AL	1	N	M	F2	AL	5	N	M	F2	AL	7	N	M	F2	AL	16
N	M	F3	AL	27	N	M	F3	AL	0	N	M	F3	AL	30	N	M	F3	AL	10
N	F	F0	SL	14	N	F	F0	SL	27	N	F	F3	AL	41	N	F	F0	AL	69
N	F	F0	AL	25	N	F	F0	SL	10	N	F	F0	SL	11	N	F	F0	AL	20
N	F	F1	SL	33	N	F	F1	SL	5	N	F	F1	AL	7	N	F	F1	SL	0
N	F	F1	AL	1	N	F	F1	SL	5	N	F	F1	SL	5	N	F	F1	SL	5
N	F	F1	AL	5	N	F	F1	AL	7	N	F	F1	SL	11	N	F	F1	SL	15
N	F	F2	SL	5	N	F	F2	SL	14	N	F	F1	AL	6	N	F	F2	AL	6
N	F	F2	SL	7	N	F	F2	SL	28	N	F	F2	AL	0	N	F	F2	SL	5
N	F	F3	SL	14	N	F	F3	SL	2	N	F	F2	SL	2	N	F	F3	SL	3
N	F	F3	AL	8	N	F	F3	AL	10	N	F	F3	AL	12	N	F	F3	AL	1
N	F	F3	AL	1	N	F	F3	AL	9	N	F	F3	AL	22	N	F	F3	AL	3
N	F	F3	AL	3	N	F	F3	AL	5	N	F	F3	AL	15	N	F	F3	AL	18
N	F	F3	AL	22	N	F	F3	AL	37										

therefore equated with a function of μ, the logit, $\eta = \log(\mu/(1-\mu))$. In generalized linear models, the linear predictor may be equated with any of a number of different functions $g(\mu)$ of μ, called *link functions*; that is,

$$\eta = g(\mu). \tag{7.2}$$

In linear regression, the probability distribution of the response variable is assumed to be normal with mean μ. In logistic regression a binomial distribution is assumed with probability parameter μ. Both distributions, the normal and Binomial distributions, come from the same family of distributions, called the exponential family,

$$f(y; \theta, \phi) = \exp\{(y\theta - b(\theta))/a(\phi) + c(y, \phi)\}. \tag{7.3}$$

For example, for the normal distribution,

$$
\begin{aligned}
f(y; \theta, \phi) &= \frac{1}{\sqrt{(2\pi\sigma^2)}} \exp\{-(y-\mu)^2/2\sigma^2\} \\
&= \exp\{(y\mu - \mu^2/2)/\sigma^2 - \frac{1}{2}(y^2/\sigma^2 + \log(2\pi\sigma^2))\} \quad (7.4)
\end{aligned}
$$

so that $\theta = \mu$, $b(\theta) = \theta^2/2$, $\phi = \sigma^2$ and $a(\phi) = \phi$

The parameter θ, a function of μ, is called the *canonical link*. The canonical link is frequently chosen as the link function (and is the default link in the Stata command for fitting generalized linear models, glm), although the canonical link is not necessarily more appropriate than any other link. Table 7.2 lists some of the most common distributions and their canonical link functions used in generalized linear models. The mean and variance of Y are given by

Table 7.2: Probability distributions and their canonical link functions

Distribution	Variance function	Dispersion parameter	Link function	$g(\mu) = \theta(\mu)$
Normal	1	σ^2	identity	μ
Binomial	$\mu(1-\mu)$	1	logit	$\log(\mu/(1-\mu))$
Poisson	μ	1	log	$\ln(\mu)$
Gamma	μ^2	ν^{-1}	reciprocal	$1/\mu$

$$E(Y) = b'(\theta) = \mu \tag{7.5}$$

and

$$\text{var}(Y) = b''(\theta)a(\phi) = V(\mu)a(\phi) \tag{7.6}$$

where $b'(\theta)$ and $b''(\theta)$ denote the fist and second derivative of $b(\theta)$ with respect to θ and the variance function $V(\mu)$ is obtained by expressing $b''(\theta)$ as a function of μ. It can be seen from (7.4) that the variance for the normal distribution is simply σ^2 regardless of the value of the mean μ, i.e., the variance function is 1.

The data on Australian school children will be analyzed by assuming a Poisson distribution for the number of days absent from school. The Poisson distribution is the appropriate distribution of the number of events observed, if these events occur independently in continuous time at a constant instantaneous probability rate (or incidence rate), see for example Clayton and Hills (1993). The Poisson distribution is given by

$$f(y;\mu) = \mu^y e^{-\mu}/y!, \quad y = 0, 1, 2, \cdots. \tag{7.7}$$

Taking the logarithm and summing over observations, the log likelihood is

$$l(\boldsymbol{\mu};\mathbf{y}) = \sum_i \{(y_i \ln \mu_i - \mu_i) - \ln(y_i!)\} \tag{7.8}$$

so that $\theta = \ln\mu$, $b(\theta) = \exp(\theta)$, $\phi = 1$, $a(\phi) = 1$ and $\operatorname{var}(y) = \exp(\theta) = \mu$. Therefore, the variance of the Poisson distribution is not constant, but equal to the mean. Unlike the normal distribution, the Poisson distribution has no separate parameter for the variance and the same is true of the Binomial distribution. Table 7.2 shows the variance functions and dispersion parameters for some commonly used probability distributions.

7.2.1 Model selection and measure of fit

Lack of fit may be expressed by the deviance, which is minus twice the difference between the maximized log likelihood of the model and the maximum likelihood achievable, i.e., the maximized likelihood of the *full* or saturated model. For the normal distribution, the deviance is simply the residual sum of squares. Another measure of lack of fit is the generalized Pearson X^2,

$$X^2 = \sum_i (y_i - \hat{\mu}_i)^2 / V(\hat{\mu}_i), \tag{7.9}$$

which, for the Poisson distribution, is just the familiar statistic for two-way cross-tabulations (since $V(\hat{\mu}) = \hat{\mu}$). Both the deviance and Pearson X^2 have χ^2 distributions when the sample size tends to infinity. When the dispersion parameter ϕ is fixed (not estimated), an analysis of deviance may be used for testing nested models in the same way as analysis of variance is used for linear models. The difference in deviance between two models is simply compared with the χ^2 distribution with degrees of freedom equal to the difference in model degrees of freedom.

The Pearson and deviance residuals are defined as the (signed) square roots of the contributions of the individual observations to the Pearson X^2 and deviance respectively. These residuals may be used to assess the appropriateness of the link and variance functions.

A relatively common phenomenon with binary and count data is *overdis-*

persion, i.e., the variance is greater than that of the assumed distribution (binomial and Poisson respectively). This overdispersion may be due to extra variability in the parameter μ which has not been completely explained by the covariates. One way of addressing the problem is to allow μ to vary randomly according to some (prior) distribution and to assume that conditional on the parameter having a certain value, the response variable follows the binomial (or Poisson) distribution. Such models are called *random effects models*.

A more pragmatic way of accommodating overdispersion in the model is to assume that the variance is proportional to the variance function, but to estimate the dispersion rather than assuming the value 1 appropriate for the distributions. For the Poisson distribution, the variance is modeled as

$$\text{var}(Y) = \phi\mu \tag{7.10}$$

where ϕ is estimated from the Deviance or Pearson X^2. (This is analogous to the estimation of the residual variance in linear regresion models from the residual sums of squares.) This parameter is then used to scale the estimated standard errors of the regression coefficients. If the variance is not proportional to the variance function, robust standard errors can be used, see next section. This approach of assuming a variance function that does not correspond to any probability distribution is an example of *quasi-likelihood*, see also Chapter 9. See McCullagh and Nelder (1989) for more details on generalized linear models.

7.2.2 Robust standard errors of parameter estimates

A very useful feature of Stata is that robust standard errors of estimated parameters can be obtained for many estimation commands. In maximum likelihood estimation, the standard errors of the estimated parameters are derived from the Hessian of the log-likelihood. However, these standard errors are correct only if the likelihood is the true likelihood of the data. If this assumption is not correct, due to misspecification of the covariates, the link function, or the probability distribution function, we can still use robust estimates of the standard errors known as the Huber, White, or sandwich variance estimates (for details, see Binder (1983)).

In the description of the robust variance estimator in the *Stata User's Guide* (Section 23.11), it is pointed out that the use of robust standard errors implies a slightly less ambitious interpretation of the parameter estimates and their standard errors than a model-based approach. The parameter estimates are unbiased estimates of the estimates that would be obtained if we had an infinite sample (not of any true parameters), and their standard errors are the standard deviations under repeated sampling followed by estimation (see also the FAQ by Sribney, 1998).

Another approach to estimating the standard errors without making any distributional assumptions is *bootstrapping* (Efron and Tibshirani, 1993). If we could obtain repeated samples from the population (from which our data were sampled), we could obtain an empirical sampling distribution of the parameter estimates. In Monte-Carlo simulation, the required samples are drawn from the assumed distribution. In bootstrapping, the sample is resampled "to approximate what would happen if the population were sampled" (Manly, 1997) . Bootstrapping works as follows. Take a random sample of n observations, with replacement, and estimate the regression coefficients. Repeat this a number of times to obtain a sample of estimates. From this sample, estimate the variance-covariance matrix of the parameter estimates. Confidence intervals may be constructed using the estimated variance or directly from the appropriate centiles of the empirical distribution of parameter estimates. See Manly (1997) and Efron and Tibshirani (1993) for more information on the bootstrap.

7.3 Analysis using Stata

The `glm` command can be used to fit generalized linear models. The syntax is analogous to `logistic` and `regress` except that the options `family()` and `link()` are used to specify the probability distribution of the response and the link function, respectively.

7.3.1 Datasets from previous chapters

First, we show how linear regression can be carried out using `glm`. In Chapter 3, the U.S. air-pollution data were read in using the instructions

```
infile str10 town so2 temp manuf pop wind precip days /*
      */ using usair.dat
drop if town=="Chicago"
```

and now we regress so2 on a number of variables using

```
glm so2 temp pop wind precip, fam(gauss) link(id)
```

```
Iteration 1 : deviance = 10150.1520

Residual df   =       35                    No. of obs =        40
Pearson X2    =  10150.15                   Deviance   =  10150.15
Dispersion    =  290.0043                   Dispersion =  290.0043

Gaussian (normal) distribution, identity link
-------------------------------------------------------------------------
   so2 |     Coef.   Std. Err.      t     P>|t|    [95% Conf. Interval]
--------+----------------------------------------------------------------
  temp | -1.810123   .4404001    -4.110   0.000   -2.704183   -.9160635
   pop |  .0113089   .0074091     1.526   0.136   -.0037323    .0263501
  wind | -3.085284   2.096471    -1.472   0.150   -7.341347    1.170778
precip |  .5660172   .2508601     2.256   0.030    .0567441    1.07529
 _cons |  131.3386   34.32034     3.827   0.001    61.66458    201.0126

(Model is ordinary regression, use regress instead)
```

The results are identical to those of the regression analysis as the remark at
the bottom of the output points out. The dispersion parameter represents the
residual variance given under **Residual MS** in the analysis of variance table of
the regression analysis in Chapter 3. We can estimate robust standard errors
using **regress** with the option **robust** (the robust estimator is not yet available
for **glm** although a Roger Newson has written a program **rglm** for this purpose,
published in STB-50 sg114):

    reg so2 temp pop wind precip, robust

```
Regression with robust standard errors       Number of obs =        40
                                             F( 4,    35) =      9.72
                                             Prob > F     =    0.0000
                                             R-squared    =    0.3446
                                             Root MSE     =     17.03

------------------------------------------------------------------------
               |          Robust
   so2 |     Coef.   Std. Err.      t     P>|t|    [95% Conf. Interval]
--------+----------------------------------------------------------------
  temp | -1.810123   .3462819    -5.227   0.000   -2.513113   -1.107134
   pop |  .0113089   .0084062     1.345   0.187   -.0057566    .0283743
  wind | -3.085284   1.792976    -1.721   0.094   -6.72522     .554651
precip |  .5660172   .1919587     2.949   0.006    .1763203    .955714
 _cons |  131.3386   19.23291     6.829   0.000    92.29368    170.3835
------------------------------------------------------------------------
```

giving slightly different standard errors which indicates that the assumption
of identically normally distributed residuals may not be entirely satisfied.
    We now show how an analysis of variance model can be fitted using **glm**,
using the slimming clinic example of chapter 5. The data are read using

    infile cond status resp using slim.dat

and the full, saturated model may be obtained using

```
xi: glm resp i.cond*i.status, fam(gauss) link(id)
```

```
i.cond              Icond_1-2   (naturally coded; Icond_1 omitted)
i.status            Istatu_1-2  (naturally coded; Istatu_1 omitted)
i.cond*i.status     IcXs_#-#    (coded as above)

Iteration 1 : deviance = 1078.8481

Residual df =        30                     No. of obs =        34
Pearson X2  =   1078.848                    Deviance    =   1078.848
Dispersion  =    35.9616                    Dispersion  =    35.9616

Gaussian (normal) distribution, identity link
------------------------------------------------------------------------
   resp |     Coef.   Std. Err.      t     P>|t|     [95% Conf. Interval]
--------+---------------------------------------------------------------
Icond_2 |   .6780002   3.234433    0.210   0.835    -5.927593    7.283594
Istatu_2|   6.128834   3.19204     1.920   0.064    -.3901823   12.64785
IcXs_2_2|  -.2655002   4.410437   -0.060   0.952    -9.272815    8.741815
  _cons |     -7.298   2.68185    -2.721   0.011    -12.77507   -1.820931
------------------------------------------------------------------------
(Model is ordinary regression, use regress instead)
```

This result is identical to that obtained using the command

```
xi: regress resp i.cond*i.status
```

(see Exercises in Chapter 5).

We can obtain the F-statistics for the interaction term by saving the deviance of the above model (residual sum of squares) in a local macro and refitting the model with the interaction removed:

```
local dev0=e(deviance)
```

```
xi: glm resp i.cond i.status, fam(gauss) link(id)
```

```
i.cond              Icond_1-2   (naturally coded; Icond_1 omitted)
i.status            Istatu_1-2  (naturally coded; Istatu_1 omitted)

Iteration 1 : deviance = 1078.9784

Residual df =        31                     No. of obs =        34
Pearson X2  =   1078.978                    Deviance    =   1078.978
Dispersion  =   34.80576                    Dispersion  =   34.80576

Gaussian (normal) distribution, identity link
------------------------------------------------------------------------
   resp |     Coef.   Std. Err.      t     P>|t|     [95% Conf. Interval]
--------+---------------------------------------------------------------
Icond_2 |   .5352102   2.163277    0.247   0.806    -3.876821    4.947242
Istatu_2|   5.989762   2.167029    2.764   0.010     1.570077   10.40945
  _cons |  -7.199832   2.094584   -3.437   0.002    -11.47176   -2.927901
------------------------------------------------------------------------
(Model is ordinary regression, use regress instead)
```

The increase in deviance caused by the removal of the interaction term represents the sum of squares of the interaction term after eliminating the main effects:

```
local dev1=e(deviance)
local ddev='dev1'-'dev0'
display 'ddev'
```

$$\boxed{.13031826}$$

and the F-statistic is simply the mean sum of squares of the interaction term after eliminating the main effects divided by the residual mean square of the full model. The numerator and denominator degrees of freedom are 1 and 30 respectively so that F and the associated p-value may be obtained as follows:

```
local f=('ddev'/1)/('dev0'/30)
display 'f'
```

$$\boxed{.00362382}$$

```
display fprob(1,30,'f')
```

$$\boxed{.95239706}$$

The general method for testing the difference in fit of two nested generalized linear models, using the difference in deviance, is not appropriate here because the dispersion parameter $\phi = \sigma^2$ was estimated.

The logistic regression analysis of chapter 6 may also be repeated using glm. We first read the data as before, without replicating records.

```
infile fr1 fr2 fr3 fr4 using tumor.dat, clear
gen therapy=int((_n-1)/2)
sort therapy
by therapy: gen sex=_n
reshape long fr, i(therapy sex) j(outc)
gen improve=outc
recode improve 1/2=0 3/4=1
list
```

	therapy	sex	outc	fr	improve
1.	0	1	1	28	0
2.	0	1	2	45	0
3.	0	1	3	29	1
4.	0	1	4	26	1
5.	0	2	1	4	0
6.	0	2	2	12	0
7.	0	2	3	5	1
8.	0	2	4	2	1
9.	1	1	1	41	0
10.	1	1	2	44	0
11.	1	1	3	20	1
12.	1	1	4	20	1
13.	1	2	1	12	0
14.	1	2	2	7	0
15.	1	2	3	3	1
16.	1	2	4	1	1

The `glm` command may be used with weights just like other estimation commands, so that we can analyze the table using the `fweight` option.

```
glm improve therapy sex [fweight=fr], fam(binomial) link(logit)
```

```
Iteration 1 : deviance =  381.8102
Iteration 2 : deviance =  381.2637
Iteration 3 : deviance =  381.2634
Iteration 4 : deviance =  381.2634

Residual df  =       296                 No. of obs =       299
Pearson X2   =  298.7045                 Deviance   =  381.2634
Dispersion   =  1.009137                 Dispersion =  1.288052

Bernoulli distribution, logit link
--------------------------------------------------------------------
improve |     Coef.   Std. Err.      z    P>|z|     [95% Conf. Interval]
--------+-----------------------------------------------------------
therapy |  -.5022014   .2456898   -2.044   0.041    -.9837444   -.0206583
    sex |  -.6543125   .3714737   -1.761   0.078    -1.382388    .0737625
  _cons |   .3858095    .451417    0.855   0.393    -.4989515    1.270571
--------------------------------------------------------------------
```

The likelihood ratio test for `sex` can be obtained as follows:

```
local dev0=e(deviance)
quietly glm improve therapy [fweight=fr], fam(binomial) link(logit)
local dev1=e(deviance)
dis 'dev1'-'dev0'
```

$$\boxed{3.3459816}$$

```
dis chiprob(1,'dev1'-'dev0')
```

$$\boxed{.0673693}$$

which gives the same result as in Chapter 6.

7.3.2 Australian school-children

We now analyze the data in Table 7.1. The data are available as a Stata file *quine.dta* and may therefore be read simply by using the command

```
use quine, clear
```

The variables are of type string and can be converted to numeric using the encode command as follows:

```
encode eth, gen(ethnic)
drop eth
encode sex, gen(gender)
drop sex
encode age, gen(class)
drop age
encode lrn, gen(slow)
drop lrn
```

The number of children in each of the combinations of categories of **gender**, **class**, and **slow** can be found using

```
table slow class ethnic, contents(freq) by(gender)
```

```
----------+-------------------------------------------------------------
          |                        ethnic and class
gender    | ----------- A ---------      ----------- N ---------
and slow  |  F0     F1    F2    F3         F0     F1    F2    F3
----------+-------------------------------------------------------------
F         |
      AL  |   4      5     1     9          4      6     1    10
      SL  |   1     10     8                1     11     9
----------+-------------------------------------------------------------
M         |
      AL  |   5      2     7     7          6      2     7     7
      SL  |   3      3     4                3      7     3
----------+-------------------------------------------------------------
```

This reveals that there were no "slow learners" in class **F3**. A table of the means and standard deviations is obtained using

```
table slow class ethnic, contents(mean days sd days) /*
    */ by(gender) format(%4.1f)
```

```
---------+------------------------------------------------
        |              ethnic and class
gender  | ---------- A ----------      ---------- N ----------
and slow|   FO    F1    F2    F3        FO    F1    F2    F3
---------+------------------------------------------------
F       |
      AL| 21.2  11.4   2.0  14.6      18.5  11.0   1.0  13.5
        | 17.7   6.5         14.9      10.7   8.9         11.5
        |
      SL|  3.0  22.6  36.4            25.0   6.0   6.2
        |       18.7  26.5                   4.2   5.0
---------+------------------------------------------------
M       |
      AL| 13.0  10.5  27.4  27.1       5.3   3.5   9.1  27.3
        |  8.0   4.9  14.7  10.4       5.4   0.7   9.5  22.9
        |
      SL|  9.0   9.0  37.0            30.0   6.1  29.3
        |  6.2   5.2  23.4            32.5   6.1   7.0
---------+------------------------------------------------
```

where the format() option causes only a single decimal place to be given. This
table suggests that the variance associated with the Poisson distribution is not
appropriate here since squaring the standard deviations (to get the variances)
results in values that are greater than the means, i.e., there is overdispersion.
In this case, the overdispersion is probably due to the fact that there is a
substantial variability in children's underlying tendency to miss days of school
that cannot be fully explained by the variables we have included in the model.

Ignoring the problem of overdispersion, a generalized linear model with a
Poisson family and a log link can be fitted using

```
glm days slow class ethnic gender, fam(pois) link(log)
```

```
Iteration 1 : deviance = 1796.1923
Iteration 2 : deviance = 1768.7236
Iteration 3 : deviance = 1768.6453
Iteration 4 : deviance = 1768.6453

Residual df  =      141              No. of obs =       146
Pearson X2   = 1990.103              Deviance    = 1768.645
Dispersion   =  14.1142              Dispersion =  12.54358

Poisson distribution, log link
-------------------------------------------------------------------------
   days |    Coef.   Std. Err.      z      P>|z|     [95% Conf. Interval]
--------+----------------------------------------------------------------
   slow | .2661578   .0445711     5.972   0.000     .1788001    .3535155
  class | .2094662   .0218242     9.598   0.000     .1666916    .2522408
 ethnic | -.5511688  .0418388   -13.174   0.000    -.6331713   -.4691663
 gender | .2256243   .0415924     5.425   0.000     .1441048    .3071438
  _cons | 2.336676   .1427907    16.364   0.000     2.056812    2.616541
-------------------------------------------------------------------------
```

The algorithm takes four iteration to converge to the maximum likelihood (or
minimum deviance) solution. In the absence of overdispersion, the dispersion

parameters based on the Pearson X^2 or the Deviance should be close to 1. The values of 14.1 and 12.5, respectively, therefore indicate that there is overdispersion. The confidence intervals are therefore likely to be too narrow. McCullagh and Nelder (1989) use the Pearson X^2 divided by the degrees of freedom to estimate the dispersion parameter for the quasi-likelihood method for Poisson models. This may be achieved using the option scale(x2):

```
glm days slow class ethnic gender, fam(pois) link(log) /*
   */ scale(x2) nolog
```

```
Residual df  =        141                      No. of obs  =        146
Pearson X2   =   1990.103                      Deviance    =   1768.645
Dispersion   =    14.1142                      Dispersion  =   12.54358

Poisson distribution, log link
----------------------------------------------------------------------------
    days |      Coef.   Std. Err.       z     P>|z|      [95% Conf. Interval]
---------+------------------------------------------------------------------
    slow |    .2661578   .1674485     1.589   0.112     -.0620352    .5943508
   class |    .2094662   .0819911     2.555   0.011      .0487667    .3701657
  ethnic |   -.5511688   .1571835    -3.507   0.000     -.8592428   -.2430948
  gender |    .2256243   .1562578     1.444   0.149     -.0806353    .531884
   _cons |    2.336676   .5364486     4.356   0.000      1.285256    3.388096
----------------------------------------------------------------------------
(Standard errors scaled using square root of Pearson X2-based dispersion)
```

Here, the option nolog was used to stop the iteration log being printed. Allowing for overdipersion has had no effect on the regression coefficients, but a large effect on the p-values and confidence intervals so that gender and slow are now no longer significant. These terms will be removed from the model. The coefficients can be interpreted as the difference in the logs of the predicted mean counts between groups. For example, the log of the predicted mean number of days absent from school for white children is -0.55 lower than that for Aboriginals.

$$\ln(\hat{\mu}_2) = \ln(\hat{\mu}_1) - 0.55 \tag{7.11}$$

Exponentiating the coefficients yields count ratios (or rate ratios). Stata exponentiates all coefficients and confidence intervals when the option eform is used:

```
glm days class ethnic, fam(pois) link(log) scale(x2) /*
   */ eform nolog
```

```
Residual df  =       143                 No. of obs =        146
Pearson X2   =  2091.267                 Deviance    =   1823.481
Dispersion   =  14.62424                 Dispersion =   12.75162

Poisson distribution, log link
-----------------------------------------------------------------------
    days |      IRR   Std. Err.       z    P>|z|     [95% Conf. Interval]
---------+-------------------------------------------------------------
   class | 1.177895   .0895243     2.154   0.031     1.014874    1.367102
  ethnic | .5782531   .0924968    -3.424   0.001     .4226303     .79118
-----------------------------------------------------------------------
(Standard errors scaled using square root of Pearson X2-based dispersion)
```

Therefore, white children are absent from school about 58% as often as Aboriginal children (95% confidence interval from 42% to 79%) after controlling for class. We have treated class as a continuous measure. To see whether this appears to be appropriate, we can form the square of class and include this in the model:

```
gen class2=class^2
glm days class class2 ethnic, fam(pois) link(log) scale(x2) /*
    */ eform nolog
```

```
Residual df  =       142                 No. of obs =        146
Pearson X2   =  2081.259                 Deviance    =   1822.56
Dispersion   =  14.65676                 Dispersion =   12.83493

Poisson distribution, log link
-----------------------------------------------------------------------
    days |      IRR   Std. Err.       z    P>|z|     [95% Conf. Interval]
---------+-------------------------------------------------------------
   class | 1.059399   .4543011     0.135   0.893     .4571295    2.455158
  class2 | 1.020512   .0825501     0.251   0.802     .8708906    1.195839
  ethnic | .5784944   .092643     -3.418   0.001     .4226525    .7917989
-----------------------------------------------------------------------
(Standard errors scaled using square root of Pearson X2-based dispersion)
```

This term is not significant so we can return to the simpler model. (Note that the interaction between class and ethnic is also not significant, see exercises.)

We now look at the residuals for this model. The post-estimation function predict that was used for regress and logistic can be used here as well. To obtain standardized Pearson residuals, use the pearson option with predict and divide the residuals by the square-root of the estimated dispersion parameter stored in e(dispersp):

```
qui glm days class ethnic, fam(pois) link(log) scale(x2)
predict resp, pearson
gen stres=resp/sqrt(e(dispersp))
```

The residuals are plotted against the linear predictor using

```
predict xb, xb
graph stres xb, xlab ylab ll("Standardized Residuals") gap(3)
```

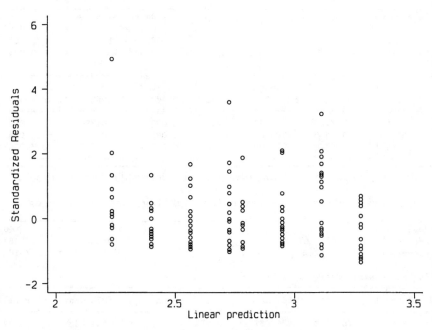

Figure 7.1 *Standardized residuals against linear predictor.*

with the result shown in Figure 7.1.

There is one large outlier. In order to find out which observation this is, we list a number of variables for cases with large standardized Pearson residuals:

```
predict mu, mu
list stres days mu ethnic class if stres>2|stres<-2
```

	stres	days	mu	ethnic	class
45.	2.03095	53	19.07713	A	F1
46.	2.09082	54	19.07713	A	F1
58.	2.070247	60	22.47085	A	F2
59.	3.228685	81	22.47085	A	F2
72.	4.924755	67	9.365361	N	F0
104.	3.588988	69	15.30538	N	F3
109.	2.019528	33	9.365361	N	F0

Case 72, a white primary school child, has a very large residual.

We now also check the assumptions of the model by estimating robust standard errors. Since **glm** does not have the **robust** option, we use the **poisson** command here.

```
poisson days class ethnic, robust nolog
```

```
Poisson regression                         Number of obs   =        146
                                           Wald chi2(2)    =      16.01
                                           Prob > chi2     =     0.0003
Log likelihood = -1205.9792                Pseudo R2       =     0.0939

-------------------------------------------------------------------------
          |               Robust
    days  |    Coef.    Std. Err.      z     P>|z|    [95% Conf. Interval]
----------+--------------------------------------------------------------
    class |  .1637288    .0766153    2.137   0.033    .0135655     .313892
   ethnic | -.5477436    .1585381   -3.455   0.001   -.8584725   -.2370147
    _cons |  3.168776    .3065466   10.337   0.000    2.567956    3.769597
-------------------------------------------------------------------------
```

Giving almost exactly the same p-values as the quasi-likelihood solution,

    glm days class ethnic, fam(pois) link(log) scale(x2) nolog

```
Residual df   =       143                No. of obs  =        146
Pearson X2    =  2091.267                Deviance    =   1823.481
Dispersion    =  14.62424                Dispersion  =   12.75162

Poisson distribution, log link
-------------------------------------------------------------------------
    days  |    Coef.    Std. Err.      z     P>|z|    [95% Conf. Interval]
----------+--------------------------------------------------------------
    class |  .1637288    .0760037    2.154   0.031    .0147643    .3126932
   ethnic | -.5477436     .159959   -3.424   0.001   -.8612575   -.2342298
    _cons |  3.168776    .3170113    9.996   0.000    2.547446    3.790107
-------------------------------------------------------------------------
(Standard errors scaled using square root of Pearson X2-based dispersion)
```

Alternatively, we can use the bootstrap command **bs**, followed by the estimation command in quotes, followed, in quotes, by expressions for the estimates for which bootstrap standard errors are required. To be on the safe side, we will ask for 500 bootstrap samples using the option **reps(500)**. First, we set the seed of the pseudorandom number generator using the **set seed** command so that we can run the sequence of commands again in the future and obtain the same results.

    set seed 12345678
    bs "poisson days class ethnic" "_b[class] _b[ethnic]", reps(500)

```
Bootstrap statistics

Variable |  Reps  Observed     Bias   Std. Err.   [95% Conf. Interval]
---------+---------------------------------------------------------------
    bs1  |  500   .1637288   .0026568  .0708082   .0246099   .3028477  (N)
         |                                        .0203793   .3041895  (P)
         |                                        .014529    .2918807  (BC)
---------+---------------------------------------------------------------
    bs2  |  500  -.5477436  -.0126206  .1462108  -.8350082  -.260479   (N)
         |                                       -.8556656  -.2822307  (P)
         |                                       -.8363549  -.2529301  (BC)
-------------------------------------------------------------------------
               N = normal, P = percentile, BC = bias-corrected
```

This compares very well with the robust Poisson and quasi-likelihood results.

We could also model overdispersion by assuming a *random effects* model where each child has an unobserved, random proneness to be absent from school. This proneness (called *frailty* in a medical context) multiplies the rate predicted by the covariates so that some children have higher or lower rates of absence from school than other children with the same covariates. The observed counts are assumed to have Poisson distribution conditional on the random effects. If the frailties are assumed to have a gamma distribution, then the (marginal) distribution of the counts has a negative binomial distribution. The negative binomial model may be fitted using **nbreg** as follows:

```
nbreg days class ethnic, nolog
```

```
Negative binomial regression                    Number of obs   =        146
                                                LR chi2(2)      =      15.77
                                                Prob > chi2     =     0.0004
Log likelihood = -551.24625                     Pseudo R2       =     0.0141

-----------------------------------------------------------------------------
     days |      Coef.   Std. Err.      z     P>|z|     [95% Conf. Interval]
----------+------------------------------------------------------------------
    class |   .1505165   .0732832     2.054   0.040     .0068841    .2941489
   ethnic |  -.5414185   .1578378    -3.430   0.001    -.8507748   -.2320622
    _cons |    3.19392   .3217681     9.926   0.000     2.563266    3.824574
----------+------------------------------------------------------------------
 /lnalpha |  -.1759664   .1243878                      -.4197619    .0678292
----------+------------------------------------------------------------------
    alpha |   .8386462   .1043173     8.039   0.000     .6572032    1.070182
-----------------------------------------------------------------------------
Likelihood ratio test of alpha=0:   chi2(1) =   1309.47   Prob > chi2 = 0.0000
```

All four methods of analyzing the data lead to the same conclusions. The Poisson model is a special case of the negative binomial model with $\alpha = 0$. The likelihood ratio test for α is therefore a test of the negative binomial against the Poisson distribution. The very small p-value "against Poisson" indicates that there is significant overdispersion.

7.4 Exercises

1. Calculate the F-statistic and difference in deviance for **status** controlling for **cond** for the data in *slim.dat*.

2. Fit the model using only **status** as the independent variable, using robust standard errors. How does this compare with a t-test with unequal variances?

3. Test the significance of the interaction between **class** and **ethnic** for the data in *quine.dta*.

4. Excluding the outlier (case 72), fit the model with explanatory variables **ethnic** and **class**.

5. Dichotomize days absent from school by classifying 14 days or more as frequently absent. Analyze this new response using both the logistic and probit link and the binomial family.

6. Use `logit` and `probit` to estimate the same models with robust standard errors and compare this with the standard errors obtained using bootstrapping.

See also the Exercises in Chapter 10.

CHAPTER 8

Analysis of Longitudinal Data I: The Treatment of Post-Natal Depression

8.1 Description of data

The data set to be analyzed in this chapter originates from a clinical trial of the use of estrogen patches in the treatment of postnatal depression; full details are given in Gregoire et al. (1996). In total, 61 one women with major depression, which began within 3 months of childbirth and persisted for up to 18 months postnatally, were allocated randomly to the active treatment or a placebo (a dummy patch); 34 received the former and the remaining 27 received the latter. The women were assessed pretreatment and monthly for six months after treatment on the Edinburgh postnatal depression scale (EPDS), higher values of which indicate increasingly severe depression. The data are shown in Table 8.1; a value of -9 in this table indicates that the observation is missing.

Table 8.1: Data in *depress.dat*

subj	group	pre	dep1	dep2	dep3	dep4	dep5	dep6
1	0	18	17	18	15	17	14	15
2	0	27	26	23	18	17	12	10
3	0	16	17	14	-9	-9	-9	-9
4	0	17	14	23	17	13	12	12
5	0	15	12	10	8	4	5	5
6	0	20	19	11.54	9	8	6.82	5.05
7	0	16	13	13	9	7	8	7
8	0	28	26	27	-9	-9	-9	-9
9	0	28	26	24	19	13.94	11	9
10	0	25	9	12	15	12	13	20
11	0	24	14	-9	-9	-9	-9	-9
12	0	16	19	13	14	23	15	11
13	0	26	13	22	-9	-9	-9	-9
14	0	21	7	13	-9	-9	-9	-9
15	0	21	18	-9	-9	-9	-9	-9
16	0	22	18	-9	-9	-9	-9	-9
17	0	26	19	13	22	12	18	13
18	0	19	19	7	8	2	5	6
19	0	22	20	15	20	17	15	13.73
20	0	16	7	8	12	10	10	12
21	0	21	19	18	16	13	16	15
22	0	20	16	21	17	21	16	18
23	0	17	15	-9	-9	-9	-9	-9
24	0	22	20	21	17	14	14	10

Table 8.1: Data in *depress.dat* (continued)

25	0	19	16	19	-9	-9	-9	-9
26	0	21	7	4	4.19	4.73	3.03	3.45
27	0	18	19	-9	-9	-9	-9	-9
28	1	21	13	12	9	9	13	6
29	1	27	8	17	15	7	5	7
30	1	15	8	12.27	10	10	6	5.96
31	1	24	14	14	13	12	18	15
32	1	15	15	16	11	14	12	8
33	1	17	9	5	3	6	0	2
34	1	20	7	7	7	12	9	6
35	1	18	8	1	1	2	0	1
36	1	28	11	7	3	2	2	2
37	1	21	7	8	6	6.5	4.64	4.97
38	1	18	8	6	4	11	7	6
39	1	27.46	22	27	24	22	24	23
40	1	19	14	12	15	12	9	6
41	1	20	13	10	7	9	11	11
42	1	16	17	26	-9	-9	-9	-9
43	1	21	19	9	9	12	5	7
44	1	23	11	7	5	8	2	3
45	1	23	16	13	-9	-9	-9	-9
46	1	24	16	15	11	11	11	11
47	1	25	20	18	16	9	10	6
48	1	22	15	17.57	12	9	8	6.5
49	1	20	7	2	1	0	0	2
50	1	20	12.13	8	6	3	2	3
51	1	25	15	24	18	15.19	13	12.32
52	1	18	17	6	2	2	0	1
53	1	26	1	18	10	13	12	10
54	1	20	27	13	9	8	4	5
55	1	17	20	10	8.89	8.49	7.02	6.79
56	1	22	12	-9	-9	-9	-9	-9
57	1	22	15.38	2	4	6	3	3
58	1	23	11	9	10	8	7	4
59	1	17	15	-9	-9	-9	-9	-9
60	1	22	7	12	15	-9	-9	-9
61	1	26	24	-9	-9	-9	-9	-9

8.2 The analysis of longitudinal data

The data in Table 8.1 consist of repeated observations over time on each of the 61 patients; such data are generally referred to as *longitudinal*. There is a large body of methods which can be used to analyze longitudinal data, ranging from the simple to the complex. Some useful references are Diggle et al. (1994), Everitt (1995), and Hand and Crowder (1996). In this chapter we shall concentrate on the following approaches:

- Graphical displays
- Summary measure or response feature analysis.

In the next chapter, more formal modeling techniques will be applied to the data.

8.3 Analysis using Stata

Assuming the data are in an ASCII file, *depress.dat*, as listed in Table 8.1, they may be read into Stata for analysis using the following instructions:

```
infile subj group pre dep1 dep2 dep3 dep4 dep5 dep6 /*
    */ using depress.dat
mvdecode _all, mv(-9)
```

The second of these instructions converts the '-9's in the data to missing values.

It is useful to begin examination of these data using the **summarize** procedure to calculate means, variances etc., within each of the two treatment groups:

```
summarize pre-dep6 if group==0
```

Variable	Obs	Mean	Std. Dev.	Min	Max
pre	27	20.77778	3.954874	15	28
dep1	27	16.48148	5.279644	7	26
dep2	22	15.88818	6.124177	4	27
dep3	17	14.12882	4.974648	4.19	22
dep4	17	12.27471	5.848791	2	23
dep5	17	11.40294	4.438702	3.03	18
dep6	17	10.89588	4.68157	3.45	20

```
summarize pre-dep6 if group==1
```

Variable	Obs	Mean	Std. Dev.	Min	Max
pre	34	21.24882	3.574432	15	28
dep1	34	13.36794	5.556373	1	27
dep2	31	11.73677	6.575079	1	27
dep3	29	9.134138	5.475564	1	24
dep4	28	8.827857	4.666653	0	22
dep5	28	7.309286	5.740988	0	24
dep6	28	6.590714	4.730158	1	23

There is a general decline in the EPDS over time in both groups, with the values in the active treatment group appearing to be consistently lower.

8.3.1 Graphical displays

A useful preliminary step in the analysis of longitudinal data is to graph the observations in some way. The aim is to highlight two particular aspects of the data, namely, how they evolve over time and how the measurements made at different times are related. A number of graphical displays can be used, including:

- separate plots of each subject's responses against time, differentiating in some way between subjects in different groups
- box plots of the observations at each time point by treatment group
- a plot of means and standard errors by treatment group for every time point
- a scatterplot matrix of the repeated measurements.

To begin, plot the required scatterplot matrix, identifying treatment groups with the labels 0 and 1, using

```
graph pre-dep6, matrix symbol([group]) ps(150)
```

The resulting plot is shown in Figure 8.1. The most obvious feature of this diagram is the increasingly strong relationship between the measurements of depression as the time interval between them decreases. This has important implications for the models appropriate for longitudinal data, as seen in Chapter 9.

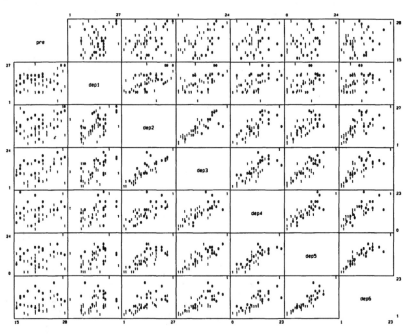

Figure 8.1 *Scatter-plot matrix for depression scores at six visits.*

To obtain the other graphs mentioned above, the data need to be restructured from its present wide form to the long form using the **reshape** command. Before running **reshape**, we will preserve the data using the **preserve** command so that they can later be restored using **restore**:

```
preserve
reshape long dep, i(subj) j(visit)
list in 1/13
```

	subj	visit	group	pre	dep
1.	1	1	0	18	17
2.	1	2	0	18	18
3.	1	3	0	18	15
4.	1	4	0	18	17
5.	1	5	0	18	14
6.	1	6	0	18	15
7.	2	1	0	27	26
8.	2	2	0	27	23
9.	2	3	0	27	18
10.	2	4	0	27	17
11.	2	5	0	27	12
12.	2	6	0	27	10
13.	3	1	0	16	17

There are a number of methods for obtaining diagrams containing separate plots of each individual's responses against time from the restructured data but the simplest is to use the connect(L) option which 'tricks' Stata into drawing lines connecting points within a specified grouping variable in a single graph. Here, the required grouping variable is subj, the y variable is dep and the x variable is visit. Before plotting, the data needs to be sorted by the grouping variable and by the x variable:

```
sort group subj visit
graph dep visit, by(group) c(L)
```

The c(L) option connects points only so long as visit is ascending. For observations with subj equal to one this is true; but for the second subject, visit begins at one again, so the last point for subject one is *not* connected with the first point for subject two. The remaining points for this subject are, however, connected and so on. Using the by(group) option with only two groups produces a display that is half empty; we therefore use the commands

```
sort subj visit
graph dep visit if group==0, c(L) ylab 11("depression") /*
    */ t1("placebo group") gap(3)
graph dep visit if group==1, c(L) ylab 11("depression") /*
    */ t1("estrogen patch group") gap(3)
```

to obtain the diagrams shown in Figure 8.2. The individual plots reflect the general decline in the depression scores over time indicated by the means obtained using the summarize command; there is, however, considerable variability. Notice that some profiles are not complete because of the presence of missing values.

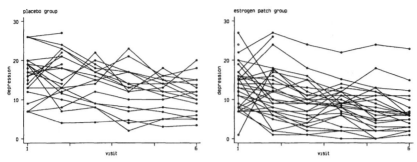

Figure 8.2 *Individual response profiles by treatment group.*

To obtain the boxplots of the depression scores at each visit for each treatment group, the following instructions can be used:

```
sort visit
graph dep if group==0, box by(visit) ll("depression")   /*
    */ ylab t1("placebo group") t2(" ") gap(3)
graph dep if group==1, box by(visit) ll("depression")   /*
    */ ylab t1("estrogen patch group") t2(" ") gap(3)
```

The resulting graphs are shown in Figure 8.3. Again, the general decline in depression scores in both treatment groups can be seen and, in the active treatment group, there is some evidence of outliers which may need to be examined. (Figure 8.2 shows that four of the outliers are due to one subject whose response profile lies above the others.)

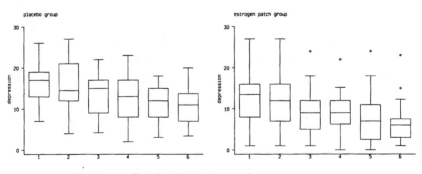

Figure 8.3 *Boxplots for six visits by treatment group.*

A plot of the mean profiles of each treatment group, which includes information about the standard errors of each mean, can be obtained using the `collapse` instruction that produces a dataset consisting of selected summary statistics. Here, we need the mean depression score on each visit for each group, the corresponding standard deviations, and a count of the number of observations on which these two statistics are based.

```
collapse (mean) dep (sd) sddep=dep (count) n=dep, by(visit group)
list in 1/10
```

	visit	group	dep	sddep	n
1.	1	0	16.48148	5.279644	27
2.	1	1	13.36794	5.556373	34
3.	2	0	15.88818	6.124177	22
4.	2	1	11.73677	6.575079	31
5.	3	0	14.12882	4.974648	17
6.	3	1	9.134138	5.475564	29
7.	4	0	12.27471	5.848791	17
8.	4	1	8.827857	4.666653	28
9.	5	0	11.40294	4.438702	17
10.	5	1	7.309286	5.740988	28

The mean value is now stored in dep; but since more than one summary statistic for the depression scores were required, the remaining statistics were given new names in the collapse instruction.

The required mean and standard error plots can now be found using

```
sort group
gen high=dep+2*sddep/sqrt(n)
gen low=dep-2*sddep/sqrt(n)
graph dep high low visit, by(group) c(lll) sort
```

Again, we can obtain better looking graphs by plotting them separately by group.

```
graph dep high low visit if group==0, c(lll) sort s(oii)    /*
    */ pen(322) yscale(5,20) ylab l1("depression") gap(3) /*
    */ t1("placebo group")
graph dep high low visit if group==1, c(lll) sort s(oii)    /*
    */ pen(322) yscale(5,20) ylab l1("depression") gap(3) /*
    */ t1("estrogen patch group")
```

Here, we have set the range of the y-axes to the same values using the option yscale() and have used the option pen() to display the confidence limits in the same colour. Before printing, we have changed the line thickness of pen 3 to 4 units (click into **Prefs** in the menu bar, select **Graph Preferences**, etc.). The resulting diagrams are shown in Figure 8.4

8.3.2 Response feature analysis

A relatively straightforward approach to the analysis of longitudinal data is that involving the use of *summary measures*, sometimes known as *response*

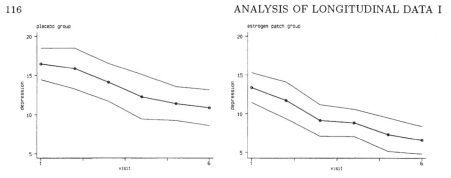

Figure 8.4 *Mean and standard error plots; the envelopes represent ± 2 standard errors.*

Table 8.2: Response features suggested in Mathews et al. (1990)

Type of data	Property to be compared between groups	Summary measure
Peaked	overall value of response	mean or area under curve
Peaked	value of most extreme response	maximum (minimum)
Peaked	delay in response	time to maximum or minimum
Growth	rate of change of response	linear regression coefficient
Growth	final level of response	final value or (relative) difference between first and last
Growth	delay in response	time to reach a particular value

feature analysis. The responses of each subject are used to construct a single number that characterizes some relevant aspect of the subject's response profile. (In some situations more than a single summary measure may be required.) The summary measure needs to be chosen prior to the analysis of the data. The most commonly used measure is the mean of the responses over time since many investigations, e.g., clinical trials, are most concerned with differences in overall levels rather than more subtle effects. Other possible summary measures are listed in Matthews et al. (1990) and are shown here in Table 8.2.

Having identified a suitable summary measure, the analysis of the data generally involves the application of a simple univariate test (usually a t-test or its nonparametric equivalent) for group differences on the single measure now available for each subject. For the estrogen patch trial data, the mean over time seems an obvious summary measure. The mean of all non-missing values is obtained (after restoring the data) using

```
restore
egen avg=rmean(dep1 dep2 dep3 dep4 dep5 dep6)
```

The differences between these means may be tested using a t-test assuming equal variances in the populations:

```
ttest avg, by(group)
```

```
Two-sample t test with equal variances

-----------------------------------------------------------------------------
  Group |    Obs       Mean    Std. Err.   Std. Dev.   [95% Conf. Interval]
--------+--------------------------------------------------------------------
      0 |     27   14.75605    .8782852    4.563704    12.95071    16.56139
      1 |     34   10.55206    .9187872    5.357404    8.682772    12.42135
--------+--------------------------------------------------------------------
combined |    61   12.41284    .6923949    5.407777    11.02785    13.79784
--------+--------------------------------------------------------------------
   diff |            4.20399   1.294842                1.613017    6.794964
-----------------------------------------------------------------------------
Degrees of freedom: 59

                 Ho: mean(0) - mean(1) = diff = 0

   Ha: diff < 0              Ha: diff ~= 0              Ha: diff > 0
     t =   3.2467              t =   3.2467              t =   3.2467
   P < t =  0.9990          P > |t| =   0.0019         P > t =   0.0010
```

Now relax the assumption of equal variances:

```
ttest avg, by(group) unequal
```

```
Two-sample t test with unequal variances

-----------------------------------------------------------------------------
  Group |    Obs       Mean    Std. Err.   Std. Dev.   [95% Conf. Interval]
--------+--------------------------------------------------------------------
      0 |     27   14.75605    .8782852    4.563704    12.95071    16.56139
      1 |     34   10.55206    .9187872    5.357404    8.682772    12.42135
--------+--------------------------------------------------------------------
combined |    61   12.41284    .6923949    5.407777    11.02785    13.79784
--------+--------------------------------------------------------------------
   diff |            4.20399   1.271045                1.660343    6.747637
-----------------------------------------------------------------------------
Satterthwaite's degrees of freedom:  58.6777

                 Ho: mean(0) - mean(1) = diff = 0

   Ha: diff < 0              Ha: diff ~= 0              Ha: diff > 0
     t =   3.3075              t =   3.3075              t =   3.3075
   P < t =  0.9992          P > |t| =   0.0016         P > t =   0.0008
```

Both tests and the associated confidence intervals indicate clearly that there is a substantial difference in overall level in the two treatment groups.

8.4 Exercises

1. How would you produce boxplots corresponding to those shown in Figure 8.3 using the data in the wide form?

2. Compare the results of the t-tests given in the text with the corresponding t-tests calculated only for those subjects having observations on all six post-randomization visits.

3. Repeat the summary measures analysis described in the text using now the mean over time divided by the standard deviation over time.

4. Test for differences in the mean over time controlling for the baseline measurement using

   (a) a change score defined as the difference between the mean over time and the baseline measurement,

   (b) analysis of covariance of the mean over time using the baseline measurement as a covariate.

   See also Exercises in Chapter 9.

CHAPTER 9

Analysis of Longitudinal Data II: Epileptic Seizures and Chemotherapy

9.1 Introduction

In a clinical trial reported by Thall and Vail (1990), 59 patients with epilepsy were randomized to groups receiving either the anti-epileptic drug progabide, or a placebo, as an adjutant to standard chemotherapy. The number of seizures was counted over four two-week periods. In addition, a baseline seizure rate was recorded for each patient, based on the eight-week prerandomization seizure count. The age of each patient was also noted. The data are shown in Table 9.1. (These data also appear in Hand et al., 1994.)

Table 9.1: Data in chemo.dat

subj	y1	y2	y3	y4	treat	base	age
1	5	3	3	3	0	11	31
2	3	5	3	3	0	11	30
3	2	4	0	5	0	6	25
4	4	4	1	4	0	8	36
5	7	18	9	21	0	66	22
6	5	2	8	7	0	27	29
7	6	4	0	2	0	12	31
8	40	20	23	12	0	52	42
9	5	6	6	5	0	23	37
10	14	13	6	0	0	10	28
11	26	12	6	22	0	52	36
12	12	6	8	4	0	33	24
13	4	4	6	2	0	18	23
14	7	9	12	14	0	42	36
15	16	24	10	9	0	87	26
16	11	0	0	5	0	50	26
17	0	0	3	3	0	18	28
18	37	29	28	29	0	111	31
19	3	5	2	5	0	18	32
20	3	0	6	7	0	20	21
21	3	4	3	4	0	12	29
22	3	4	3	4	0	9	21
23	2	3	3	5	0	17	32

Table 9.1: Data in chemo.dat (continued)

24	8	12	2	8	0	28	25
25	18	24	76	25	0	55	30
26	2	1	2	1	0	9	40
27	3	1	4	2	0	10	19
28	13	15	13	12	0	47	22
29	11	14	9	8	1	76	18
30	8	7	9	4	1	38	32
31	0	4	3	0	1	19	20
32	3	6	1	3	1	10	30
33	2	6	7	4	1	19	18
34	4	3	1	3	1	24	24
35	22	17	19	16	1	31	30
36	5	4	7	4	1	14	35
37	2	4	0	4	1	11	27
38	3	7	7	7	1	67	20
39	4	18	2	5	1	41	22
40	2	1	1	0	1	7	28
41	0	2	4	0	1	22	23
42	5	4	0	3	1	13	40
43	11	14	25	15	1	46	33
44	10	5	3	8	1	36	21
45	19	7	6	7	1	38	35
46	1	1	2	3	1	7	25
47	6	10	8	8	1	36	26
48	2	1	0	0	1	11	25
49	102	65	72	63	1	151	22
50	4	3	2	4	1	22	32
51	8	6	5	7	1	41	25
52	1	3	1	5	1	32	35
53	18	11	28	13	1	56	21
54	6	3	4	0	1	24	41
55	3	5	4	3	1	16	32
56	1	23	19	8	1	22	26
57	2	3	0	1	1	25	21
58	0	0	0	0	1	13	36
59	1	4	3	2	1	12	37

9.2 Possible models

The data listed in Table 9.1 consist of repeated observations on the same subject taken over time, and are a further example of a set of *longitudinal data*. During the last decade, statisticians have considerably enriched the methodology available for the analysis of such data (see Diggle, Liang, and Zeger, 1994) and many of these developments are implemented in Stata.

Models for the analysis of longitudinal data are similar to the generalized linear models encountered in Chapter 7, but with one very important difference, namely, the residual terms are allowed to be correlated rather than independent. This is necessary since the observations at different time points in a longitudinal study involve the same subjects, thus generating some pattern of dependence which needs to be accounted for by any proposed model.

9.2.1 Normally distributed responses

If we suppose that a normally distributed response is observed on each individual at T time points, then the basic regression model for longitudinal data becomes (cf. equation (3.2))

$$\mathbf{y}_i = \mathbf{X}\boldsymbol{\beta} + \boldsymbol{\epsilon}_i \tag{9.1}$$

where $\mathbf{y}_i' = (y_{i1}, y_{i2}, \cdots, y_{iT})$, $\boldsymbol{\epsilon}' = (\epsilon_{i1}, \cdots, \epsilon_{iT})$, $\mathbf{X}$ is a $T \times (p+1)$ design matrix, and $\boldsymbol{\beta}' = (\beta_0, \cdots, \beta_p)$ is a vector of parameters. Assuming the residual terms have a multivariate normal distribution with a particular covariance matrix allows maximum likelihood estimation to be used; details are given in Jennrich and Schluchter (1986). If all covariance parameters are estimated independently, giving an unstructured covariance matrix, then this approach is essentially equivalent to multivariate analysis of variance for longitudinal data.

Another approach is to try to explain the covariances by introducing latent variables, the simplest example being a *random intercept model* given by

$$y_{ij} = \boldsymbol{\beta}^T \mathbf{x}_{ij} + u_i + \epsilon_{ij} \tag{9.2}$$

where the unobserved random effects u_i and the residuals ϵ_{ij} are assumed to be independently normally distributed with zero means and constant variances. This random effects model implies that the covariances of the responses y_{ij} and y_{ik} at different time points are all equal to each other and that the variances at each time point are constant, a structure of the covariance matrix known as *compound symmetry* (for example, see Winer, 1971). If compound symmetry is assumed, this is essentially equivalent to assuming a split-plot design.

Other correlation structures include autoregressive models where the correlations decrease with the distance between time-points and the off-diagonals of the correlation matrix a constant. Whatever the assumed correlation structure, all models may be estimated by maximum likelihood.

9.2.2 Non-normal responses

In cases where normality cannot be assumed, it is not possible to specify a likelihood with an arbitrary correlation structure. We can define random effects models by introducing a random intercept into the linear predictor,

$$\eta_{ij} = \boldsymbol{\beta}^T \mathbf{x}_{ij} + u_i \tag{9.3}$$

where the u_i are independently distributed. (The negative binomial model is an example of the model above where there is only one observation per subject, see Chapter 7.) The attraction of such models, also known as generalized linear mixed models, is that they correspond to a probabilistic mechanism that may have generated the data and that estimation is via maximum likelihood. However, generalized linear mixed models tend to be difficult to estimate (for example, see Goldstein, 1995) and the implied correlation structures (of random intercept models) are not sufficiently general for all purposes.

In the generalized estimating equation approach introduced by Liang, and Zeger (1986), any required covariance structure and any link function may be assumed and parameters estimated without specifying the joint distribution of the repeated observations. Estimation is via a quasi-likelihood approach (see Wedderburn, 1974). This is analogous to the approach taken in Chapter 7 where we freely estimated the dispersion ϕ for a Poisson model that is characterized by $\phi = 1$.

Since the parameters specifying the structure of the correlation matrix are rarely of great practical interest (they are what is known as *nuisance parameters*), simple structures are used for the within-subject correlations giving rise to the so-called *working correlation matrix*. Liang and Zeger (1986) show that the estimates of the parameters of most interest, i.e., those that determine the mean profiles over time, are still valid even when the correlation structure is incorrectly specified.

The two approaches – random effects modeling and generalized estimating equations – lead to different interpretations of between subject effects. In random effects models, a between-subject effect represents the difference between subjects conditional on having the same random effect, whereas the parameters of generalized estimating equations represent the average difference between subjects. The two types of model are therefore also known as *conditional* and *marginal* models, respectively. In practice, this distinction is important only if link functions other than the identity or log link are used, for example in logistic regression (see Diggle et al., 1994).

A further issue with many longitudinal data sets is the occurrence of dropouts, i.e., subjects who fail to complete all scheduled visits. (The depression data of the previous chapter suffered from this problem.) A taxonomy of dropouts is given in Diggle, Liang and Zeger (1994) where it is shown that it is necessary to make particular assumptions about the dropout mechanism for the analyses described in this chapter to be valid.

9.3 Analysis using Stata

We will first describe the generalized estimating equations approach to modeling longitudinal data which has been implemented in the **xtgee** command of Stata. The main components of a model which have to be specified are:

- the assumed distribution of the response variable, specified in the `family()` option;
- the link between the response variable and its linear predictor, specified in the `link()` option; and
- the structure of the working correlation matrix, specified in the `correlations()` option.

In general, it is not necessary to specify both `family()` and `link()` since, as explained in Chapter 7, the default link is the canonical link for the specified family.

The `xtgee` command will often be used with the `family(gauss)` option, together with the identity link function, giving rise to multivariate normal regression, the multivariate analoge of multiple regression as described in Chapter 3. We will illustrate this option on the post-natal depression data used in the previous chapter.

9.3.1 Post-natal depression data

The data are obtained using the instructions

```
infile subj group pre dep1 dep2 dep3 dep4 dep5 dep6 /*
    */ using depress.dat
reshape long dep, i(subj) j(visit)
mvdecode _all, mv(-9)
```

To begin, fit a model that regresses depression on `group`, `pre`, and `visit` under the unrealistic assumptions of independence. The necessary instruction written out in its fullest form is

```
xtgee dep group pre visit, i(subj) t(visit) corr(indep) /*
    */ link(iden) fam(gauss)
```

(see Display 9.1)

Here, the fitted model is simply the multiple regression model described in Chapter 3 for 295 observations which are assumed to be independent of one another; the scale parameter is equal to the residual mean square and the deviance is equal to the residual sum of squares. The estimated regression coefficients and their associated standard errors indicate that the covariates `group`, `pre`, and `visit` are all significant. However, treating the observations as independent is unrealistic and will almost certainly lead to poor estimates of the standard errors. Standard errors for between subject factors (here `group` and `pre`) are likely to be underestimated because we are treating observations from the same subject as independent, thus increasing the apparent sample size; standard errors for within subject factors (here `visit`) are likely to be overestimated.

```
Iteration 1: tolerance = 1.174e-14

GEE population-averaged model            Number of obs       =      295
Group variable:                   subj   Number of groups    =       61
Link:                         identity   Obs per group: min  =        1
Family:                       Gaussian                  avg  =      4.8
Correlation:               independent                  max  =        6
                                         Wald chi2(3)        =   144.15
Scale parameter:               25.80052  Prob > chi2         =   0.0000

Pearson chi2(291):              7507.95  Deviance            =  7507.95
Dispersion (Pearson):          25.80052  Dispersion          = 25.80052

-----------------------------------------------------------------------
    dep |      Coef.   Std. Err.      z    P>|z|     [95% Conf. Interval]
--------+--------------------------------------------------------------
  group |  -4.290664   .6072954    -7.065  0.000    -5.480941   -3.100387
    pre |   .4769071   .0798565     5.972  0.000     .3203913    .633423
  visit |  -1.307841    .169842    -7.700  0.000    -1.640725   -.9749569
  _cons |   8.233577   1.803945     4.564  0.000     4.697909   11.76924
-----------------------------------------------------------------------
```

Display 9.1

Now abandon the assumption of independence and estimate a correlation matrix having compound symmetry (i.e., constrain the correlations between the observations at any pair of time points to be equal). Such a correlational structure is introduced using corr(exchangeable), or the abbreviated form corr(exc). The model can be fitted as follows:

```
xtgee dep group pre visit, i(subj) t(visit) corr(exc) /*
*/ link(iden) fam(gauss)
```

Instead of specifying the subject and time identifiers using the options i() and t(), we can also declare the data as being of the form xt (for cross-sectional time series) as follows:

```
iis subj
tis visit
```

and omit the i() and t() options from now on. Since both the link and the family correspond to the default options, the same analysis may be carried out using the shorter command

```
xtgee dep group pre visit, corr(exc)
```

(see Display 9.2)

After estimation, xtcorr reports the working correlation matrix.

```
xtcorr
```

```
Iteration 1: tolerance = .04955537
Iteration 2: tolerance = .00043788
Iteration 3: tolerance = 4.514e-06
Iteration 4: tolerance = 4.659e-08

GEE population-averaged model            Number of obs      =      295
Group variable:                    subj  Number of groups   =       61
Link:                          identity  Obs per group: min =        1
Family:                        Gaussian                 avg =      4.8
Correlation:                exchangeable                 max =        6
                                         Wald chi2(3)       =   132.47
Scale parameter:               25.91572  Prob > chi2        =   0.0000

------------------------------------------------------------------------------
     dep |     Coef.  Std. Err.       z    P>|z|    [95% Conf. Interval]
---------+--------------------------------------------------------------------
   group | -4.026258   1.085886   -3.708   0.000   -6.154556   -1.897961
     pre |  .4599716   .1447738    3.177   0.001    .1762202    .743723
   visit | -1.227231   .1188545  -10.325   0.000   -1.460182   -.9942806
   _cons |  8.432182   3.134731    2.690   0.007    2.288222    14.57614
------------------------------------------------------------------------------
```

Display 9.2

```
Estimated within-subj correlation matrix R:

         c1      c2      c3      c4      c5      c6
r1   1.0000
r2   0.5510  1.0000
r3   0.5510  0.5510  1.0000
r4   0.5510  0.5510  0.5510  1.0000
r5   0.5510  0.5510  0.5510  0.5510  1.0000
r6   0.5510  0.5510  0.5510  0.5510  0.5510  1.0000
```

Note that the standard errors for **group** and **pre** have increased whereas that for **visit** has decreased as expected. The estimated within subject correlation matrix demonstrates the compound symmetry structure.

As we pointed out before, compound symmetry arises when a random subject effect is introduced into the linear model for the observations. Such a random effects model can be fitted by restricted maximum likelihood using **xtreg**

```
xtreg dep group pre visit, i(subj)
```

(see Display 9.3) giving almost the same parameter estimates as the GEE model. We also obtain the variance components, with the between subject standard deviation, **sigma_u**, estimated as 3.78 and the within-subject standard deviation, **sigma_e**, estimated as 3.36. The correlation between time-points predicted by this model is the ratio of the between-subject variance to the total variance and may be calculated using

```
disp 3.777959^2/ 5.053135^2
```

```
.55897538
```

```
Random-effects GLS regression            Number of obs      =        295
Group variable (i) : subj                Number of groups   =         61

R-sq:  within  = 0.2984                   Obs per group: min =          1
       between = 0.3879                                  avg =        4.8
       overall = 0.3312                                  max =          6

Random effects u_i ~ Gaussian            Wald chi2(3)       =     134.33
corr(u_i, X)       = 0 (assumed)         Prob > chi2        =     0.0000

-----------------------------------------------------------------------------
     dep |      Coef.   Std. Err.      z     P>|z|     [95% Conf. Interval]
---------+-------------------------------------------------------------------
   group | -4.023373   1.089009    -3.695   0.000    -6.157792   -1.888954
     pre |  .4598446   .1452149     3.167   0.002     .1752288    .7444605
   visit |  -1.22638   .1176654   -10.423   0.000       -1.457   -.9957599
   _cons |  8.433315   3.143717     2.683   0.007     2.271742   14.59489
---------+-------------------------------------------------------------------
 sigma_u | 3.7779589
 sigma_e | 3.3557717
     rho |  .55897526   (fraction of variance due to u_i)
-----------------------------------------------------------------------------
```

Display 9.3

(given under rho in the output) which is also close to the GEE estimate. The above model was fitted using *restricted maximum likelihood* (REML). Use option `mle` to obtain the maximum likelihood solution.

The compound symmetry structure for the correlations implied by the random effects model is frequently not acceptable since it is more likely that correlations between pairs of observations widely separated in time will be lower than for observations made closer together. This pattern was apparent from the scatterplot matrix given in the previous chapter.

To allow for a more complex pattern of correlations among the repeated observations, we can move to an *autoregressive structure*. For example, in a first order autoregressive specification the correlation between time points r and s is assumed to be $\rho^{|r-s|}$. The necessary instruction for fitting the previously considered model but with this first order autoregressive structure for the correlations is

```
xtgee dep group pre visit, corr(ar1)
```

The output shown in Display 9.4 includes the note that eight subjects had to be excluded because they had fewer than two observations so that they could not contribute to the estimation of the correlation matrix. The estimates of the regression coefficients and their standard errors have changed but not substantially. The estimated within-subject correlation matrix may again be obtained using

```
xtcorr
```

```
note:  some groups have fewer than 2 observations
       not possible to estimate correlations for those groups
       8 groups omitted from estimation


Iteration 1: tolerance = .10123769
Iteration 2: tolerance = .00138587
Iteration 3: tolerance = .00002804
Iteration 4: tolerance = 5.704e-07

GEE population-averaged model          Number of obs     =      287
Group and time vars:       subj visit  Number of groups  =       53
Link:                       identity   Obs per group: min =        2
Family:                     Gaussian                 avg =      5.4
Correlation:                  AR(1)                   max =        6
                                       Wald chi2(3)      =    63.59
Scale parameter:            26.19077   Prob > chi2       =   0.0000

------------------------------------------------------------------------
    dep |    Coef.   Std. Err.      z    P>|z|    [95% Conf. Interval]
--------+---------------------------------------------------------------
  group | -4.217426  1.062939   -3.968   0.000   -6.300747  -2.134104
    pre |  .4265161  .1388479    3.072   0.002    .1543792   .6986531
  visit | -1.181364  .1918328   -6.158   0.000   -1.55735   -.805379
  _cons |  9.042893  3.062883    2.952   0.003    3.039752   15.04603
------------------------------------------------------------------------
```

Display 9.4

```
Estimated within-subj correlation matrix R:

        c1       c2       c3       c4       c5       c6
r1  1.0000
r2  0.6835  1.0000
r3  0.4671  0.6835  1.0000
r4  0.3192  0.4671  0.6835  1.0000
r5  0.2182  0.3192  0.4671  0.6835  1.0000
r6  0.1491  0.2182  0.3192  0.4671  0.6835  1.0000
```

which has the expected pattern in which correlations decrease substantially as the separation between the observations increases.

Other correlation structures are available using **xtgee**, including the option **correlation(unstructured)** in which no constraints are placed on the correlations (This is essentially equivalent to multivariate analysis of variance for longitudinal data, except that the variance is assumed to be constant.) It might appear that using this option would be the most sensible one to choose for *all* data sets. This is not, however, the case since it necessitates the estimation of many nuisance parameters. This can, in some circumstances, cause problems in the estimation of those parameters of most interest, particularly when the sample size is small and the number of time points is large.

9.3.2 Chemotherapy data

We now analyze the chemotherapy data using a similar model as for the depression data, again using baseline, group, and time variables as covariates, but using the Poisson distribution and log link. Assuming the data are available in an ASCII file called *chemo.dat*, they may be read into Stata in the usual way using

```
infile subj y1 y2 y3 y4 treat baseline age using chemo.dat
```

Some useful summary statistics can be obtained using

```
summarize y1 y2 y3 y4 if treat==0
```

Variable	Obs	Mean	Std. Dev.	Min	Max
y1	28	9.357143	10.13689	0	40
y2	28	8.285714	8.164318	0	29
y3	28	8.785714	14.67262	0	76
y4	28	7.964286	7.627835	0	29

```
summarize y1 y2 y3 y4 if treat==1
```

Variable	Obs	Mean	Std. Dev.	Min	Max
y1	31	8.580645	18.24057	0	102
y2	31	8.419355	11.85966	0	65
y3	31	8.129032	13.89422	0	72
y4	31	6.709677	11.26408	0	63

The largest value of **y1** in the progabide group seems out of step with the other maximum values and may indicate an outlier. Some graphics of the data may be useful for investigating this possibility further, but first it is convenient to transform the data from its present 'wide' form to the 'long' form. The necessary command is as follows:

```
reshape long y, i(subj) j(week)
sort subj treat week
list in 1/12
```

	subj	week	treat	baseline	age	y
1.	1	1	0	11	31	5
2.	1	2	0	11	31	3
3.	1	3	0	11	31	3
4.	1	4	0	11	31	3
5.	2	1	0	11	30	3
6.	2	2	0	11	30	5
7.	2	3	0	11	30	3
8.	2	4	0	11	30	3
9.	3	1	0	6	25	2
10.	3	2	0	6	25	4
11.	3	3	0	6	25	0
12.	3	4	0	6	25	5

Perhaps the most useful graphical display for investigating the data is a set of graphs of individual response profiles. To include the baseline value in the graphs, we duplicate the first observation for each subject (week=1) and change one of the duplicates for each subject to represent the baseline measure (week=0). Since the baseline measure represents seizure counts over an eight-week period, compared with two-week periods for each of the other time-points, we divide the baseline measure by 4:

```
expand 2 if week==1
sort subj week
qui by subj: replace week=0 if _n==1
replace y=baseline/4 if week==0
```

Since we are planning to fit a Poisson model with the log link to the data, we take the log transformation before plotting the response profiles. (We need to add 1 because some seizure counts are zero.)

```
gen ly = log(y+1)
```

To avoid too much overlap between subjects, we use nine graphs in each group. First we produce a grouping variable dum9 that splits each treatment group into nine groups. If subjects are numbered by a variable i, we can apply the function mod(i,9) to create a variable that numbers subjects from $0, \cdots, 8, 0 \cdots, 8, \cdots$. The graphs are clearer to read if the baseline values within a graph are not too similar. We therefore sort by baseline within each treatment group and define i to number subjects in this order.

```
sort treat baseline subj week
quietly by treat baseline subj: replace i= _n==1
replace i=sum(i)
gen dum9=mod(i,9)
```

We can now produce the graphs:

```
sort dum9 subj week
graph ly week if treat==0, c(L) s([subj]) by(dum9) /*
```

```
     */ 11(" ") bs(5) b1("placebo group") b2(" ")
graph ly week if treat==1, c(L) s([subj]) by(dum9) /*
     */ 11(" ") bs(5) b1("progabide group") b2(" ")
drop if week==0
```

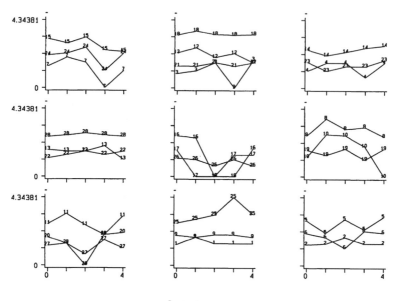

placebo group

Figure 9.1 *Response profiles in placebo group.*

The resulting graphs are shown in Figure 9.1 and 9.2. There is no obvious improvement in the progabide group. Subject 49 had more epileptic fits overall than any other subject. However, since we are going to control for the baseline measure, it is an unusual shape (not the overall level) of the response profile that makes a subject an outlier.

As discussed in Chapter 7, the most plausible distribution for count data is the Poisson distribution. The Poisson distribution is specified in xtgee models using the option family(poisson). The log link is implied (since it is the canonical link). The summary tables for the seizure data given above provide strong empirical evidence that there is overdispersion (the variances are greater than the means) and this can be incorporated using the scale(x2) option to allow the dispersion parameter ϕ to be estimated (see also Chapter 7).

```
iis subj
xtgee y treat baseline age week, corr(exc) family(pois) /*
     */ scale(x2)
```

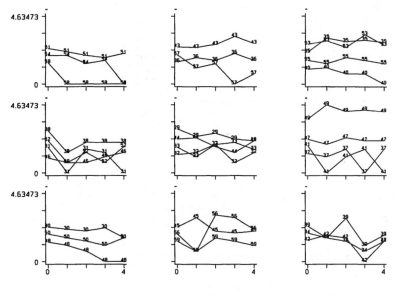

progabide group

Figure 9.2 *Response profiles in the treated group.*

The output assuming an exchangeable correlation structure is given in Display 9.5 and the estimated correlation matrix is obtained using **xtcorr**.

xtcorr

```
Estimated within-subj correlation matrix R:

        c1       c2       c3       c4
r1  1.0000
r2  0.4004  1.0000
r3  0.4004  0.4004  1.0000
r4  0.4004  0.4004  0.4004  1.0000
```

In Display 9.5, the parameter ϕ is estimated as 5.11, indicating the severe over-dispersion in these data. We briefly illustrate how important it was to allow for overdispersion by omitting the **scale(x2)** option:

xtgee y treat baseline age week, corr(exc) family(pois)

The results given in Display 9.6 show that the variable **treat** becomes significant leading to a qualitatively different conclusion about the effectiveness of the treatment. Even if overdispersion had not been suspected, this error could have been detected by using the **robust** option (see Chapter 7):

```
Iteration 1: tolerance = .01808267
Iteration 2: tolerance = 2.553e-06
Iteration 3: tolerance = 9.247e-10

GEE population-averaged model          Number of obs      =       236
Group variable:                  subj  Number of groups   =        59
Link:                             log  Obs per group: min =         4
Family:                       Poisson                 avg =       4.0
Correlation:              exchangeable                 max =         4
                                       Wald chi2(4)       =    190.68
Scale parameter:              5.107826  Prob > chi2       =    0.0000
```

y	Coef.	Std. Err.	z	P>|z|	[95% Conf. Interval]
treat	-.1479036	.1600935	-0.924	0.356	-.4616812	.1658739
baseline	.022742	.0017046	13.341	0.000	.019401	.026083
age	.0235616	.0134748	1.749	0.080	-.0028486	.0499718
week	-.0587233	.035548	-1.652	0.099	-.1283962	.0109495
_cons	.6763124	.4622316	1.463	0.143	-.2296448	1.58227

(Standard errors scaled using square root of Pearson X2-based dispersion)

Display 9.5

```
Iteration 1: tolerance = .01808267
Iteration 2: tolerance = 2.553e-06
Iteration 3: tolerance = 9.246e-10

GEE population-averaged model          Number of obs      =       236
Group variable:                  subj  Number of groups   =        59
Link:                             log  Obs per group: min =         4
Family:                       Poisson                 avg =       4.0
Correlation:              exchangeable                 max =         4
                                       Wald chi2(4)       =    973.97
Scale parameter:                     1  Prob > chi2       =    0.0000
```

y	Coef.	Std. Err.	z	P>|z|	[95% Conf. Interval]
treat	-.1479036	.0708363	-2.088	0.037	-.2867402	-.0090671
baseline	.022742	.0007542	30.152	0.000	.0212637	.0242203
age	.0235616	.0059622	3.952	0.000	.011876	.0352473
week	-.0587233	.0157289	-3.733	0.000	-.0895514	-.0278953
_cons	.6763124	.2045227	3.307	0.001	.2754553	1.07717

Display 9.6

```
xtgee y treat baseline age week, corr(exc) family(pois) robust
```

```
Iteration 1: tolerance = .01808267
Iteration 2: tolerance = 2.553e-06
Iteration 3: tolerance = 9.247e-10

GEE population-averaged model          Number of obs       =        236
Group variable:                  subj  Number of groups    =         59
Link:                             log  Obs per group: min  =          4
Family:                       Poisson                 avg  =        4.0
Correlation:              exchangeable                 max  =          4
                                       Wald chi2(4)        =     603.44
Scale parameter:                    1  Prob > chi2         =     0.0000

                                (standard errors adjusted for clustering on subj)
------------------------------------------------------------------------------
             |               Semi-robust
         y   |      Coef.   Std. Err.      z    P>|z|     [95% Conf. Interval]
-------------+----------------------------------------------------------------
       treat | -.1479036    .1701506   -0.869   0.385    -.4813928    .1855855
    baseline |   .022742    .0012542   18.133   0.000     .0202839    .0252001
         age |  .0235616    .0119013    1.980   0.048     .0002355    .0468877
        week | -.0587233     .035298   -1.664   0.096    -.1279062    .0104595
       _cons |  .6763124    .3570204    1.894   0.058    -.0234346    1.376059
------------------------------------------------------------------------------
```

Display 9.7

The results of the robust regression in Display 9.7 are remarkably similar to those of the overdispersed Poisson model, indicating that the latter is a reasonable model for the data.

The estimated coefficient of **treat** describes the difference in the log of the average seizure counts between the placebo and progabide treated groups. The negative value indicates that the treatment is more effective than the placebo in controlling the seizure rate although this is not significant. The exponentiated coefficient gives an incidence rate ratio; here it represents the ratio of average seizure rates, measured as the number of seizures per two-week period, for the treated patients compared to that among the control patients. The exponentiated coefficient and the corresponding confidence interval can be obtained directly using the **eform** option in **xtgee**:

```
xtgee y treat baseline age week, corr(exc) /*
*/ scale(x2) family(pois) eform
```

The results in Display 9.8 indicate that there is a 14% reduction in the incidence rate of epileptic seizures in the treated group compared with the control group. According to the confidence interval, the reduction could be as low as 37% or there could be a much as an 18% increase.

```
Iteration 1: tolerance = .01808267
Iteration 2: tolerance = 2.553e-06
Iteration 3: tolerance = 9.247e-10


GEE population-averaged model              Number of obs       =       236
Group variable:                     subj   Number of groups    =        59
Link:                                log   Obs per group: min  =         4
Family:                          Poisson                  avg  =       4.0
Correlation:                 exchangeable                 max  =         4
                                           Wald chi2(4)        =    190.68
Scale parameter:                5.107826   Prob > chi2         =    0.0000


------------------------------------------------------------------------------
       y |      IRR    Std. Err.      z     P>|z|     [95% Conf. Interval]
---------+--------------------------------------------------------------------
   treat | .8625142    .1380829    -0.924   0.356     .6302232    1.180424
baseline | 1.023003    .0017438    13.341   0.000      1.01959    1.026426
     age | 1.023841    .0137961     1.749   0.080     .9971555    1.051241
    week | .9429676    .0335206    -1.652   0.099     .8795048     1.01101
------------------------------------------------------------------------------
(Standard errors scaled using square root of Pearson X2-based dispersion)
```

Display 9.8

When these data were analyzed by Thall and Vail (1990), a possible interaction between baseline seizure count and treatment group was allowed for. Such a model is easily fitted using the following instructions:

```
gen blint = treat*baseline
xtgee y treat baseline age week blint, /*
    */corr(exc) scale(x2) family(pois) eform
```

There is no evidence of an interaction in Display 9.9. Other correlation structures might be explored for these data (see Exercises).

We now look at the standardized Pearson residuals of the previous model separately for each week, as an informal method for finding outliers (see equation (7.9)). This can be done using the **predict** command to obtain estimated counts and computing the standardized Pearson residuals as follows:

```
quietly xtgee y treat baseline age week, corr(exc) /*
    */ family(pois) scale(x2)
predict pred, xb
replace pred = exp(pred)
gen pres = (y-pred)/sqrt(pred)
gen stpres = pres/sqrt(e(chi2_dis))
sort week
graph stpres, box s([subj]) by(week) ylab
```

The resulting graph is shown in Figure 9.3

```
Iteration 1: tolerance = .01714003
Iteration 2: tolerance = .00001041
Iteration 3: tolerance = 1.088e-08

GEE population-averaged model            Number of obs      =        236
Group variable:                  subj    Number of groups   =         59
Link:                             log    Obs per group: min =          4
Family:                       Poisson                   avg =        4.0
Correlation:              exchangeable                   max =          4
                                         Wald chi2(5)       =     191.71
Scale parameter:              5.151072   Prob > chi2        =     0.0000

------------------------------------------------------------------------
      y |      IRR   Std. Err.      z    P>|z|    [95% Conf. Interval]
--------+---------------------------------------------------------------
  treat | .7735759   .1991683   -0.997   0.319    .4670343    1.281318
baseline| 1.021598   .0031501    6.930   0.000    1.015443    1.027791
    age | 1.025588   .0142413    1.820   0.069    .9980523    1.053884
   week | .9429676   .0336645   -1.645   0.100    .8792419    1.011312
  blint | 1.001998    .003713    0.539   0.590     .994747    1.009302
------------------------------------------------------------------------
(Standard errors scaled using square root of Pearson X2-based dispersion)
```

Display 9.9

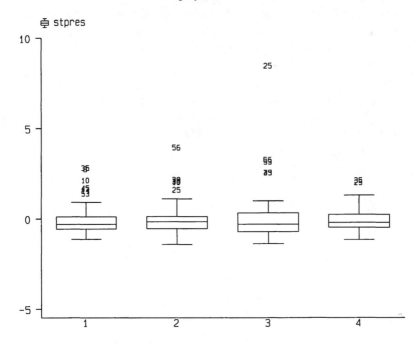

Figure 9.3 *Standardized Pearson residuals.*

Here, subject 25 is shown to be an outlier in week 3. In Figure 9.1, it is apparent that this is due to a large increase in the log of the seizure counts at week 3.

9.4 Exercises

1. For the depression data, run the random effects model using different estimation methods available for **xtreg**. (See the *Stata Reference Manual* under 'xtreg' for the interpretation of the **fe** and **be** options.)

2. Compare the result of the above with the result of running ordinary linear regression with standard errors corrected for the within-subject correlation using:

   (a) the option, **robust, cluster(subj)**, see help for **regress**, and

   (b) bootstrapping, by sampling subjects with replacement, not observations. This may be achieved using the **bs** command, together with the option **cluster(subj)**.

3. Explore other possible correlational structures for the seizure data in the context of a Poisson model. Examine the robust standard errors in each case.

4. Investigate what Poisson models are most suitable when subject 25 is excluded from the analysis.

5. Repeat the analysis treating the baseline measure (divided by four) as another response variable and replacing the variable **week** by a simple indicator variable, **post**, which is 0 at baseline and 1 subsequently. Is there a significant treatment effect, now represented by the interaction between **post** and **treat**?

CHAPTER 10

Some Epidemiology

10.1 Description of data

This chapter illustrates a number of different problems in epidemiology using four datasets that are presented in the form of cross-tabulations in Table 10.1. The first table (a) is taken from Clayton and Hills (1993). The data result from a cohort study which was carried out to investigate the relationship between energy intake and ischaemic heart disease (IHD). Low energy intake indicates lack of physical exercise and is therefore considered a risk factor. The table gives frequencies of IHD by ten-year age-band and exposure to a high or low calory diet. The total person-years of observation are also given for each cell.

The second dataset (b) is the result of a case-control study investigating whether keeping a pet bird is a risk factor for lung cancer. This dataset is given in Hand et al. (1994).

The last two datasets (c) and (d) are from matched case control studies. In dataset (c), cases of endomitral cancer were matched on age, race, date of admission, and hospital of admission to a suitable control not suffering from cancer. Past exposure to conjugated estrogens was determined. The dataset is described in Everitt (1994). The last set of data, described in Clayton and Hills (1993), arise from a case-control study of breast cancer screening. Women who had died of breast cancer were matched with three control women. The screening history of the subjects in each matched case-control set was assessed over the period up to the time of diagnosis of the case.

10.2 Introduction to epidemiology

Epidemiology is the study of diseases in populations, in particular the search for causes of disease. For ethical reasons, subjects cannot be randomized to possible risk factors in order to establish whether these are associated with an increase in the incidence of disease. Instead, epidemiology is based on observational studies, the most important types being cohort studies and case-control studies. We will give a very brief description of the design and analysis of these two types of studies, following closely the explanations and notation given in the excellent book, *Statistical Models in Epidemiology* by Clayton and Hills (1993).

In a cohort study, a group of subjects free of the disease is followed up and the presence of risk factors as well as the occurrence of the disease of interest are

Table 10.1 Simple tables in epidemiology. (a) number of IHD cases and person-years of observation by age and exposure to low energy diet; (b) number of lung cancer cases and controls who keep a pet bird; (c) frequency of exposure to oral conjugated estrogens among cases of endomitral cancer and their matched controls; (d) screening history in subjects who died of breast cancer and 3 matched controls. (Tables (a) and (d) taken from Clayton and Hills (1993) with permission of their publisher, Oxford University Press.)

(a) Cohort Study

Age	Exposed < 2750 kcal		Unexposed > 2750 kcal	
	Cases	Pers-yrs	Cases	Pers-yrs
40-49	2	311.9	4	607.9
50-59	12	878.1	5	1272.1
60-69	14	667.5	8	888.9

(b) Case-Control Study

Kept pet birds	Cases	Controls
yes	98	101
no	141	328
Total	239	429

(c) 1:1 Matched Case-Control Study

		Controls		
		+	-	Total
Cases	+	12	43	55
	-	7	121	128
	Total	19	164	183

(d) 1:3 Matched Case-Control Study

	Number of controls screened			
Status of the case	0	1	2	3
Screened	1	4	3	1
Unscreened	11	10	12	4

recorded. This design is illustrated in Figure 10.1. An example of a cohort study is the study described in the previous section where subjects were followed up to monitor the occurrence of ischaemic heart disease in two risk groups, those with high and low energy intake, giving the results in Table 10.1(a). The incidence rate λ (of the disease) may be estimated by the number of new cases of the disease D during a time interval divided by the person-time of observation Y, the sum of all subjects' periods of observation during the time interval:

$$\hat{\lambda} = \frac{D}{Y}. \tag{10.1}$$

This is the maximum likelihood estimator of λ assuming that D follows a Poisson distribution (independent events occurring at a constant probability rate in continuous time) with mean λY. The most important quantity of interest in a cohort study is the incidence rate ratio (or relative risk) between subgroups of subjects defined by their exposure to the risk factor. This may be estimated by

$$\hat{\theta} = \frac{D_1/Y_1}{D_0/Y_0} \tag{10.2}$$

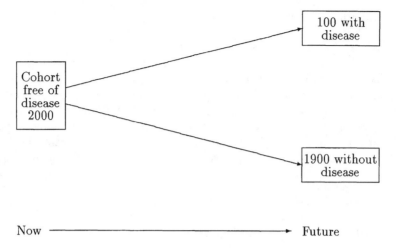

Figure 10.1 *Cohort study.*

where the subscripts 1 and 0 denote exposed and unexposed groups, respectively. This estimator may be derived by maximizing the conditional likelihood that there were D_1 cases in the exposed group conditional on there being a total of $D = D_0 + D_1$ cases. Since this conditional likelihood for θ is a true likelihood based on the binomial distribution, exact confidence intervals may be obtained. However, a potential problem in estimating this rate ratio is confounding arising from systematic differences in prognostic factors between the groups. This problem can be dealt with by dividing the cohort into groups or strata according to such prognostic factors and assuming that the rate ratio for exposed and unexposed subjects is the same in each stratum. If there are D^t cases and Y^t person-years of observation in stratum t, then the common rate ratio may be estimated using the method of Mantel and Haenszel as follows

$$\hat{\theta} = \frac{\sum_t D_1^t Y_0^t / Y^t}{\sum_t D_0^t Y_1^t / Y^t}. \tag{10.3}$$

Note that the strata might not correspond to groups of subjects. For example, if the confounder is age, subjects who cross from one age-band into the next during the study, contribute parts of their periods of observation to different strata. This is how Table 10.1(a) was constructed. Another way of controlling for confounding variables is to use Poisson regression to model the number of failures. If a log link is used, the expected number of failures can be made proportional to the person-years of observation by adding the log of the person-years of observation to the linear predictor as an offset (a term that is added to the linear predictor without estimating a regression coefficient for it), giving

$$\log(D) = \log(Y) + \boldsymbol{\beta}^T \mathbf{x}. \tag{10.4}$$

Exponentiating the equation gives

$$D = Y \exp(\beta^T \mathbf{x}) \qquad (10.5)$$

as required.

If the incidence rate of a disease is small, a cohort study requires a large number of person-years of observation making it very expensive. A more feasible type of study in this situation is a case-control study in which cases of the disease of interest are compared with non-cases, called controls, with respect to exposure to possible risk factors in the past. A diagram illustrating the basic idea of case-control studies is shown in Figure 10.2. The assumption

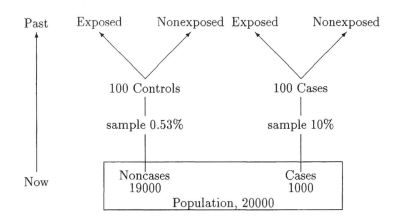

Figure 10.2 *Case-Control Study.*

here is that the probability of selection into the study is independent of the exposures of interest. The data in Table 10.1(b) derive from a case-control study in which cases with lung cancer and healthy controls were interviewed to ascertain whether they had been "exposed" to a pet bird.

Let D and H be the number of cases and controls, respectively, and let the subscripts 0 and 1 denote "unexposed" and "exposed." Since the proportion of cases was determined by the design, it is not possible to estimate the relative risk of disease between exposed and nonexposed subjects. However, the odds of exposure in the cases or controls may be estimated and the ratio of these odds

$$\frac{D_1/D_0}{H_1/H_0} = \frac{D_1/H_1}{D_0/H_0} \qquad (10.6)$$

is equal to the odds ratio of being a case in the exposed group compared with the unexposed group. If we model the (log) odds of being a case using logistic regression with the exposure as an explanatory variable, then the coefficient of the exposure variable is an estimate of the true log odds ratio even though the estimate of the odds itself (reflected by the constant) only reflects the proportion of cases in the study. Logistic regression may therefore be used to control for confounding variables.

A major difficulty with case-control studies is to find suitable controls who are similar enough to the cases (so that differences in exposure can resonably be assumed to be due their association with the disease) without being over-matched, which can result in very similar exposure patterns. The problem of finding cases who are similar enough is often addressed by matching cases individually to controls according to important variables such as age and sex. Examples of such matched case-control studies are given in Table 10.1(b) and (c). In the screening study, matching had the following additional advantage noted in Clayton and Hills (1993). The screening history of controls could be determined by considering only the period up to the diagnosis of the case, ensuring that cases did not have a decreased opportunity for screening because they would not have been screened after their diagnosis. The statistical analysis has to take account of the matching. Two methods of analysis are McNemar's test in the case of two by two tables and conditional logistic regression in the case of several controls per case and/or several explanatory variables.

Since the case-control sets have been matched on variables that are believed to be associated with disease status, the sets can be thought of as strata with subjects in one stratum having higher or lower odds of being a case than those in another after controlling for the exposures. A logistic regression model would have to accommodate these differences by including a parameter ω_c for each case-control set so that the log odds of being a case for subject i in case-control set c would be

$$\log(\omega_{ci}) = \log(\omega_c) + \beta^T \mathbf{x}_i. \tag{10.7}$$

However, this would result in too many parameters to be estimated. Furthermore, the parameters ω_c are of no interest to us. In conditional logistic regression, these nuisance parameters are eliminated as follows. In a 1:1 matched case-control study, ignoring the fact that each set has one case, the probability that subject 1 in the set is a case and subject 2 is a noncase is

$$\Pr(1) = \frac{\omega_{c1}}{1 + \omega_{c1}} \times \frac{1}{1 + \omega_{c2}} \tag{10.8}$$

and the probability that subject 1 is a noncase and subject 2 is a case is

$$\Pr(2) = \frac{1}{1 + \omega_{c1}} \times \frac{\omega_{c2}}{1 + \omega_{c2}}. \tag{10.9}$$

However, conditional on there being one case in a set, the probability of subject 1 being the case is simply

$$\frac{\Pr(1)}{\Pr(1) + \Pr(2)} = \omega_{c1} / (\omega_{c1} + \omega_{c2}) = \frac{\exp(\beta^T \mathbf{x}_1)}{\exp(\beta^T \mathbf{x}_1) + \exp(\beta^T \mathbf{x}_2)} \tag{10.10}$$

since ω_c cancels out, see equation (10.7). The expression on the right-hand side of equation (10.10) is the contribution of a single case-control set to the conditional likelihood of the sample. Similarly, it can be shown that if there are k controls per case and the subjects within each case-control set are labelled 1 for the case and 2 to $k + 1$ for the controls then the log-likelihood becomes

$$\sum_c \log \left(\frac{\exp(\boldsymbol{\beta}^T \mathbf{x}_1)}{\sum_{i=2}^{k+1} \exp(\boldsymbol{\beta}^T \mathbf{x}_i)} \right).$$ (10.11)

10.3 Analysis using Stata

There is a collection of instructions in Stata, the `epitab` commands, that may be used to analyze small tables in epidemiology. These commands either refer to variables in an existing dataset or can take cell counts as arguments (i.e., there are immediate commands, see Chapter 1).

The first dataset (a) is given in a file as tabulated and may be read using

```
infile str5 age num0 py0 num1 py1 using ihd.dat,clear
gen agegr=_n
reshape long num py, i(agegr) j(exposed)
```

Ignoring age, the incidence rate ratio may be estimated using

```
ir num exposed py
```

```
                    | exposed                    |
                    | Exposed    Unexposed  |       Total
--------------------+------------------------+-----------
              num |      17           28 |          45
               py |    2769         1857 |        4626
--------------------+------------------------+-----------
                    |                            |
   Incidence Rate |  .0061394     .0150781 |    .0097276
                    |                            |
                    |   Point estimate     | [95% Conf. Interval]
                    |------------------------+
  Inc. rate diff. |      -.0089387        | -.0152401    -.0026372
 Inc. rate ratio |       .4071738        |  .2090942     .7703146  (exact)
  Prev. frac. ex. |       .5928262        |  .2296854     .7909058  (exact)
 Prev. frac. pop |       .3548499        |
                    +---------------------------------------------
                      (midp)    Pr(k<=17) =                     0.0016  (exact)
                      (midp)  2*Pr(k<=17) =                     0.0031  (exact)
```

The incidence rate ratio is estimated as 0.41 with a 95% confidence interval from 0.21 to 0.77. The terms (exact) imply that that the confidence intervals are exact (no approximation was used). Controlling for age using the `epitab` command

```
ir num exposed py, by(age)
```

```
        age |      IRR     [95% Conf. Interval]    M-H Weight
------------+--------------------------------------------------
      40-49 |   1.026316    .1470886   11.3452      1.321739    (exact)
      50-59 |   .2876048    .0793725    .8770737    7.099535    (exact)
      60-69 |   .4293749    .1560738   1.096694     7.993577    (exact)
------------+--------------------------------------------------
      Crude |   .4071738    .2090942    .7703146                (exact)
M-H combined|   .4161246    .2262261    .7654276
------------+--------------------------------------------------
Test for heterogeneity (M-H)   chi2(2) =    1.57  Pr>chi2 = 0.4554
```

gives very similar estimates as shown in the row labeled "M-H combined" (the Mantel-Haentzel estimate). Another way of controlling for age is to carry out Poisson regression with the log of **py** as an offset. The offset may be specified using the **exposure(py)** option:

```
poisson num exposed, exposure(py) irr
```

```
Iteration 0:   log likelihood = -13.906416
Iteration 1:   log likelihood = -13.906416

Poisson regression                         Number of obs    =          6
                                           LR chi2(1)       =       8.89
                                           Prob > chi2      =     0.0029
Log likelihood = -13.906416                Pseudo R2        =     0.2422

------------------------------------------------------------------------
     num |      IRR   Std. Err.      z     P>|z|    [95% Conf. Interval]
---------+--------------------------------------------------------------
 exposed |  .4072982   .125232    -2.921   0.003    .2229428    .7441002
      py | (exposure)
------------------------------------------------------------------------
```

showing that there is an incidence rate ratio of 0.41 with a 95% confidence interval from 0.22 to 0.74. Another way of achieving the same result is using **glm** with the **offset** option which requires the logarithm of the person-years to be computed.

```
gen lpy=log(py)
glm num exposed, fam(poisson) link(log) offset(lpy) eform
```

where the **eform** option is used to obtain exponentiated coefficients.

We will analyze the case-control study using the "immediate" command **cci**. The following notation is used for **cci**

	Exposed	Unexposed
Cases	a	b
Noncases	c	d

where the quantities a, b, etc. in the table are used in alphabetical order, i.e.,

```
cci a b c d
```

(See `help epitab` for the arguments required for other immediate epitab commands.) The bird data may therefore be analyzed using the immediate command for case-control studies, `cci`, as follows:

```
cci 98 141 101 328
```

```
                                                      Proportion
                   |  Exposed   Unexposed  |    Total    Exposed
-------------------+-----------------------+----------------------
           Cases  |      98         141    |      239     0.4100
        Controls  |     101         328    |      429     0.2354
-------------------+-----------------------+----------------------
           Total  |     199         469    |      668     0.2979
                  |                        |
                  |    Pt. Est.            |   [95% Conf. Interval]
                  |------------------------+----------------------
      Odds ratio  |     2.257145           |   1.606121    3.172163  (Cornfield)
   Attr. frac. ex. |     .5569624          |    .3773821    .6847577  (Cornfield)
   Attr. frac. pop |     .2283779          |
                  +-------------------------------------------
                       chi2(1) =    22.37  Pr>chi2 = 0.0000
```

There is therefore a significant association between keeping pet birds and developing lung cancer. The word (**Cornfield**) is there to remind us that an approximation, the Cornfield approximation, was used to estimate the confidence interval (for example, see Rothman, 1986).

The matched case-control study with one control per case may be analyzed using the immediate command `mcci` where the columns of the two-way crosstabulation represent exposed and unexposed controls:

```
mcci 12 43 7 121
```

```
                |  Controls                |
Cases           |  Exposed    Unexposed    |     Total
----------------+--------------------------+----------
      Exposed   |     12           43      |       55
    Unexposed   |      7          121      |      128
----------------+--------------------------+----------
        Total   |     19          164      |      183

McNemar's chi2(1) =      25.92          Pr>chi2 = 0.0000
Exact McNemar significance probability        = 0.0000

Proportion with factor
        Cases       .3005464
        Controls    .1038251          [95% conf. interval]
                    ---------          --------------------
        difference  .1967213           .1210924    .2723502
        ratio       2.894737          1.885462    4.444269
        rel. diff.  .2195122           .1448549    .2941695

        odds ratio  6.142857          2.739803    16.18481    (exact)
```

suggesting an increased odds of endimetrial cancer in subjects exposed to oral conjugated estrogens.

The matched case-control study with three controls per case cannot be analyzed using epitab. Instead, we will use conditional logistic regression. We need to convert the data from the table shown in 10.1(d) into the form required for conditional logistic regression; that is, one observation per subject (including cases and controls); an indicator variable, cancer, for cases; another indicator variable, screen, for screening and a third variable, caseid, giving the id for each case-control set of four women.

First, read the data and transpose them so that the columns in the transposed dataset have variable names ncases0 and ncases1. The rows in this dataset correspond to different numbers of matched controls who have been screened, which will be called nconstr. Then reshape to long, generating an indicator, casescr for whether or not the case was screened:

```
infile v1-v4 using screen.dat,clear
gen str8 _varname="ncases1" in 1
replace _varname="ncases0" in 2
xpose, clear
gen nconscr=_n-1
reshape long ncases, i(nconscr) j(casescr)
list
```

	nconscr	casescr	ncases
1.	0	0	11
2.	0	1	1
3.	1	0	10
4.	1	1	4
5.	2	0	12
6.	2	1	3
7.	3	0	4
8.	3	1	1

The next step is to replicate each of the records ncases times so that we have one record per case-control set. Then define the variable caseid and expand the dataset four times in order to have one record per subject. The four subjects within each case-control are arbitrarily labeled 0 to 3 in the variable control where 0 stands for "the case" and 1, 2, and 3 for the controls.

```
expand ncases
sort casescr nconscr
gen caseid=_n
expand 4
sort caseid
quietly by caseid: gen control=_n-1
list in 1/8
```

	nconscr	casescr	ncases	caseid	control
1.	0	0	11	1	0
2.	0	0	11	1	1
3.	0	0	11	1	2
4.	0	0	11	1	3
5.	0	0	11	2	0
6.	0	0	11	2	1
7.	0	0	11	2	2
8.	0	0	11	2	3

Now **screen**, the indicator whether a subject was screened, is defined to be zero except for the cases who were screened and for as many controls as were screened according to **nconscr**. The variable **cancer** is one for cases and zero otherwise.

```
gen screen=0
replace screen=1 if control==0&casescr==1  /* the case */
replace screen=1 if control==1&nconscr>0
replace screen=1 if control==2&nconscr>1
replace screen=1 if control==3&nconscr>2
gen cancer=cond(control==0,1,0)
```

We can reproduce Table 10.1(d) by temporarily collapsing the data (using **preserve** and **restore** to revert back to the original data) as follows:

```
preserve
collapse (sum) screen (mean) casescr , by(caseid)
gen nconscr=screen-casescr
tabulate casescr nconscr
restore
```

(mean)		nconscr			
casescr	0	1	2	3	Total
0	11	10	12	4	37
1	1	4	3	1	9
Total	12	14	15	5	46

And now we are ready to carry out conditional logistic regression:

```
clogit cancer screen, group(caseid) or
```

```
Iteration 0:   Log Likelihood =-62.404527
Iteration 1:   Log Likelihood =-59.212727
Iteration 2:   Log Likelihood = -59.18163
Iteration 3:   Log Likelihood =-59.181616

Conditional (fixed-effects) logistic regression        Number of obs =     184
                                                       chi2(1)       =    9.18
                                                       Prob > chi2   = 0.0025
Log Likelihood = -59.181616                            Pseudo R2     = 0.0719

-------------------------------------------------------------------------------
  cancer | Odds Ratio   Std. Err.        z     P>|z|      [95% Conf. Interval]
---------+---------------------------------------------------------------------
  screen |   .2995581    .1278367    -2.825    0.005       .1297866     .6914043
-------------------------------------------------------------------------------
```

Screening is therefore protective of death from breast cancer, reducing the odds to a third (95% 0.13 to 0.69).

10.4 Exercises

1. Carry out conditional logistic regression to estimate the odds ratio for the data in Table 10.1(c). The data are given in the same form as in the table in a file called *estrogen.dat*.

2. For the data *ihd.dat,* use the command iri to calculate the incidence rate ratio for IHD without controlling for age.

3. Use Poisson regression to test whether the effect of exposure on incidence of IHD differs significantly between age groups.

4. Repeat the analysis of Exercise 5 in Chapter 9, using the raw seizure counts (without dividing the baseline count by 4) as dependent variables and the log of the observation time (8 weeks for baseline and 2 weeks for all other counts) as an offset.

CHAPTER 11

Survival Analysis: Retention of Heroin Addicts in Methadone Maintenance Treatment

11.1 Description of data

The "survival data" to be analyzed in this chapter are the times that heroin addicts remained in a clinic for methadone maintenance treatment. The data are given in Table 11.1. Here the endpoint of interest is not death as the word "survival" suggests, but termination of treatment. Some subjects were still in the clinic at the time these data were recorded and this is indicated by the variable **status**, which is equal to one if the person departed and 0 otherwise. Possible explanatory variables for retention in treatment are maximum methadone dose and a prison record as well as which of two clinics the addict was treated in. These variables are called **dose**, **prison**, and **clinic**, respectively. The data were first analyzed by Caplehorn and Bell (1991) and also appear in Hand et al. (1994).

Table 11.1: Data in *heroin.dat*

id	clinic	status	time	prison	dose	id	clinic	status	time	prison	dose
1	1	1	428	0	50	132	2	0	633	0	70
2	1	1	275	1	55	133	2	1	661	0	40
3	1	1	262	0	55	134	2	1	232	1	70
4	1	1	183	0	30	135	2	1	13	1	60
5	1	1	259	1	65	137	2	0	563	0	70
6	1	1	714	0	55	138	2	0	969	0	80
7	1	1	438	1	65	143	2	0	1052	0	80
8	1	0	796	1	60	144	2	0	944	1	80
9	1	1	892	0	50	145	2	0	881	0	80
10	1	1	393	1	65	146	2	1	190	1	50
11	1	0	161	1	80	148	2	1	79	0	40
12	1	1	836	1	60	149	2	0	884	1	50
13	1	1	523	0	55	150	2	1	170	0	40
14	1	1	612	0	70	153	2	1	286	0	45
15	1	1	212	1	60	156	2	0	358	0	60
16	1	1	399	1	60	158	2	0	326	1	60
17	1	1	771	1	75	159	2	0	769	1	40
18	1	1	514	1	80	160	2	1	161	0	40
19	1	1	512	0	80	161	2	0	564	1	80
21	1	1	624	1	80	162	2	1	268	1	70
22	1	1	209	1	60	163	2	0	611	1	40
23	1	1	341	1	60	164	2	1	322	0	55
24	1	1	299	0	55	165	2	0	1076	1	80
25	1	0	826	0	80	166	2	0	2	1	40
26	1	1	262	1	65	168	2	0	788	0	70

Table 11.1: Data in *heroin.dat* (continued)

27	1	0	566	1	45	169	2	0	575	0	80
28	1	1	368	1	55	170	2	1	109	1	70
30	1	1	302	1	50	171	2	0	730	1	80
31	1	0	602	0	60	172	2	0	790	0	90
32	1	1	652	0	80	173	2	0	456	1	70
33	1	1	293	0	65	175	2	1	231	1	60
34	1	0	564	0	60	176	2	1	143	1	70
36	1	1	394	1	55	177	2	0	86	1	40
37	1	1	755	1	65	178	2	0	1021	0	80
38	1	1	591	0	55	179	2	0	684	1	80
39	1	0	787	0	80	180	2	1	878	1	60
40	1	1	739	0	60	181	2	1	216	0	100
41	1	1	550	1	60	182	2	0	808	0	60
42	1	1	837	0	60	183	2	1	268	1	40
43	1	1	612	0	65	184	2	0	222	0	40
44	1	0	581	0	70	186	2	0	683	0	100
45	1	1	523	0	60	187	2	0	496	0	40
46	1	1	504	1	60	188	2	1	389	0	55
48	1	1	785	1	80	189	1	1	126	1	75
49	1	1	774	1	65	190	1	1	17	1	40
50	1	1	560	0	65	192	1	1	350	0	60
51	1	1	160	0	35	193	2	0	531	1	65
52	1	1	482	0	30	194	1	0	317	1	50
53	1	1	518	0	65	195	1	0	461	1	75
54	1	1	683	0	50	196	1	1	37	0	60
55	1	1	147	0	65	197	1	1	167	1	55
57	1	1	563	1	70	198	1	1	358	0	45
58	1	1	646	1	60	199	1	1	49	0	60
59	1	1	899	0	60	200	1	1	457	1	40
60	1	1	857	0	60	201	1	1	127	0	20
61	1	1	180	1	70	202	1	1	7	1	40
62	1	1	452	0	60	203	1	1	29	1	60
63	1	1	760	0	60	204	1	1	62	0	40
64	1	1	496	0	65	205	1	0	150	1	60
65	1	1	258	1	40	206	1	1	223	1	40
66	1	1	181	1	60	207	1	0	129	1	40
67	1	1	386	0	60	208	1	0	204	1	65
68	1	0	439	0	80	209	1	1	129	1	50
69	1	0	563	0	75	210	1	1	581	0	65
70	1	1	337	0	65	211	1	1	176	0	55
71	1	0	613	1	60	212	1	1	30	0	60
72	1	1	192	1	80	213	1	1	41	0	60
73	1	0	405	0	80	214	1	0	543	0	40
74	1	1	667	0	50	215	1	0	210	1	50
75	1	0	905	0	80	216	1	1	193	1	70
76	1	1	247	0	70	217	1	1	434	0	55
77	1	1	821	0	80	218	1	1	367	0	45
78	1	1	821	1	75	219	1	1	348	1	60
79	1	0	517	0	45	220	1	0	28	0	50
80	1	0	346	1	60	221	1	0	337	0	40
81	1	1	294	0	65	222	1	0	175	1	60
82	1	1	244	1	60	223	2	1	149	1	80
83	1	1	95	1	60	224	1	1	546	1	50
84	1	1	376	1	55	225	1	1	84	0	45
85	1	1	212	0	40	226	1	0	283	1	80
86	1	1	96	0	70	227	1	1	533	0	55
87	1	1	532	0	80	228	1	1	207	1	50
88	1	1	522	1	70	229	1	1	216	0	50
89	1	1	679	0	35	230	1	0	28	0	50
90	1	0	408	0	50	231	1	1	67	1	50
91	1	0	840	0	80	232	1	0	62	1	60
92	1	0	148	1	65	233	1	0	111	0	55

Table 11.1: Data in *heroin. dat* (continued)

93	1	1	168	0	65	234	1	1	257	1	60
94	1	1	489	0	80	235	1	1	136	1	55
95	1	0	541	0	80	236	1	0	342	0	60
96	1	1	205	0	50	237	2	1	41	0	40
97	1	0	475	1	75	238	2	0	531	1	45
98	1	1	237	0	45	239	1	0	98	0	40
99	1	1	517	0	70	240	1	1	145	1	55
100	1	1	749	0	70	241	1	1	50	0	50
101	1	1	150	1	80	242	1	0	53	0	50
102	1	1	465	0	65	243	1	0	103	1	50
103	2	1	708	1	60	244	1	0	2	1	60
104	2	0	713	0	50	245	1	1	157	1	60
105	2	0	146	0	50	246	1	1	75	1	55
106	2	1	450	0	55	247	1	1	19	1	40
109	2	0	555	0	80	248	1	1	35	0	60
110	2	1	460	0	50	249	2	0	394	1	80
111	2	0	53	1	60	250	1	1	117	0	40
113	2	1	122	1	60	251	1	1	175	1	60
114	2	1	35	1	40	252	1	1	180	1	60
118	2	0	532	0	70	253	1	1	314	0	70
119	2	0	684	0	65	254	1	0	480	0	50
120	2	0	769	1	70	255	1	0	325	1	60
121	2	0	591	0	70	256	2	1	280	0	90
122	2	0	769	1	40	257	1	1	204	0	50
123	2	0	609	1	100	258	2	1	366	0	55
124	2	0	932	1	80	259	2	0	531	1	50
125	2	0	932	1	80	260	1	1	59	1	45
126	2	0	587	0	110	261	1	1	33	1	60
127	2	1	26	0	40	262	2	1	540	0	80
128	2	0	72	1	40	263	2	0	551	0	65
129	2	0	641	0	70	264	1	1	90	0	40
131	2	0	367	0	70	266	1	1	47	0	45

The main reason why survival data require special methods of analysis is because they often contain right censored observations; that is, observations for which the endpoint of interest has not occurred during the period of observation; all that is known about the true survival time is that it exceeds the period of observation. Even if there are no censored observations, survival times tend to have positively skewed distributions that can be difficult to model.

11.2 Describing survival times and Cox's regression model

The survival time T may be regarded as a random variable with a probability distribution $F(t)$ and probability density function $f(t)$. Then an obvious function of interest is the probability of surviving to time t or beyond, the *survivor function* or survival curve $S(t)$, which is given by

$$S(t) = P(T \geq t) = 1 - F(t). \qquad (11.1)$$

A further function which is of interest for survival data is the *hazard function*; this represents the instantaneous failure rate, that is, the probability that an individual experiences the event of interest at a time point given that the event has not yet occurred. It can be shown that the hazard function is given by

$$h(t) = \frac{f(t)}{S(t)}, \tag{11.2}$$

the instantaneous probability of failure at time t divided by the probability of surviving up to time t. Note that the hazard function is just the incidence rate discussed in Chapter 10. It follows from equations (11.1) and (11.2) that

$$\frac{-d\log(S(t))}{dt} = h(t) \tag{11.3}$$

so that

$$S(t) = \exp(-H(t)) \tag{11.4}$$

where $H(t)$ is the integrated hazard function, also known as the *cumulative hazard* function.

11.2.1 Cox's regression

Cox's regression is a semi-parametric approach to survival analysis. The method does not require the probability distribution $F(t)$ to be specified; however, unlike most nonparametric methods, Cox's regression does use regression parameters in the same way as generalized linear models. The model can be written as

$$h(t) = h_0(t)\exp(\beta^T \mathbf{x}) \tag{11.5}$$

so that the hazard functions of any two individuals are assumed to be constant multiples of each other, the multiple being $\exp(\beta^T(\mathbf{x}_i - \mathbf{x}_j))$, the *hazard ratio* or incidence rate ratio. The assumption of a constant hazard ratio is called the *proportional hazards* assumption. The set of parameters $h_0(t)$, called the baseline hazard function, can be thought of as nuisance parameters whose purpose is merely to control the parameters of interest β for any changes in the hazard over time. The parameters β are estimated by maximizing the *partial log likelihood* given by

$$\sum_f \log\left(\frac{\exp(\beta^T \mathbf{x}_f)}{\sum_{i\in r(f)}\exp(\beta^T \mathbf{x}_i)}\right) \tag{11.6}$$

where the first summation is over all failures f and the second summation is over all subjects $r(f)$ still alive (and therefore "at risk") at the time of failure. It can be shown that this log-likelihood is a log profile likelihood (i.e., the log of the likelihood in which the nuisance parameters have been replaced by functions of β which maximise the likelihood for fixed β). Note also that the likelihood in equation (11.6) is equivalent to the likelihood for matched case-control studies described in Chapter 10 if the subjects at risk at the time of a failure (the *risk set*) are regarded as controls matched to the case failing at that point in time (see Clayton and Hills, 1993).

The baseline hazards may be estimated by maximizing the full log likelihood with the regression parameters evaluated at their estimated values. These hazards are nonzero only when a failure occurs. Integrating the hazard function gives the cumulative hazard function

$$H(t) = H_0(t) \exp(\boldsymbol{\beta}^T \mathbf{x}) \tag{11.7}$$

where $H_0(t)$ is the integral of $h_0(t)$. The survival curve may be obtained from $H(t)$ using equation (11.4).

It follows from equation (11.4), that the survival curve for a Cox model is given by

$$S(t) = S_0(t)^{\exp(\boldsymbol{\beta}^T \mathbf{x})} \tag{11.8}$$

The log of the cumulative hazard function predicted by the Cox model is given by

$$\log(H(t)) = \log H_0(t) + \boldsymbol{\beta}^T \mathbf{x}, \tag{11.9}$$

so that the log cumulative hazard functions of any two subjects i and j are parallel with constant difference given by $\boldsymbol{\beta}^T (\mathbf{x}_i - \mathbf{x}_j)$.

If the subjects fall into different groups and we are not sure whether we can make the assumption that the group's hazard functions are proportional to each other, we can estimate separate log cumulative hazard functions for the groups using a stratified Cox model. These curves may then be plotted to assess whether they are sufficiently parallel. For a stratified Cox model, the partial likelihood has the same form as in equation (11.6) except that the risk set for a failure is now confined to subjects in the same stratum.

Survival analysis is described in detail in Collett (1994) and in Clayton and Hills (1993).

11.3 Analysis using Stata

The data are available as an ASCII file called *heroin.dat* on the disk accompanying Hand et al. (1994). Since the data are stored in a two-column format with the set of variables repeated twice in each row, as shown in Table 11.1, we have to use reshape to bring the data into the usual form:

```
infile id1 clinic1 status1 time1 prison1 dose1 /*
    */ id2 clinic2 status2 time2 prison2 dose2 /*
    */ using heroin.dat
gen row=_n
reshape long id clinic status time prison dose, /*
    */ i(row) j(col)
drop row col
```

Before fitting any survival time models, we declare the data as being of the form "st" (for survival time) using the stset command

```
stset time, failure(status)
```

```
      failure event:  status ~= 0 & status ~= .
 obs. time interval:  (0, time]
  exit on or before:  failure

 -----------------------------------------------------------------------
      238  total obs.
        0  exclusions
 -----------------------------------------------------------------------
      238  obs. remaining, representing
      150  failures in single record/single failure data
    95812  total analysis time at risk, at risk from t =         0
                                 earliest observed entry t =         0
                                  last observed exit t =         1076
```

and look at a summary of the data using

```
stsum
```

```
      failure _d:  status
 analysis time _t:  time

        |                  incidence     no. of    |------ Survival time -----|
        |  time at risk      rate       subjects     25%       50%        75%
 -------+----------------------------------------------------------------------
  total |      95812       .0015656         238       212       504        821
```

There are therefore 238 subjects with a median survival time of 504 days. If
the incidence rate (i.e., the hazard function) could be assumed to be constant,
it would be estimated as 0.0016 per day which corresponds to 0.57 per year.

The data come from two different clinics and it is likely that these clinics have
different hazard functions which may well not be parallel. A Cox regression
model with clinics as strata and the other two variables, dose and prison as
explanatory variables is fitted using the stcox command with the strata()
option:

```
stcox dose prison, strata(clinic)
```

giving the output shown in Display 11.1. Therefore, subjects with a prison history are 47.5% more likely to drop out at any given time (given that they remained until that time) than those without a prison history. For every increase
in methadone dose by one unit (1mg), the hazard is multiplied by 0.965. This
coefficient is very close to one, but this may be because one unit of methadone
dose is not a large quantity. Even if we know little about methadone maintenance treatment, we can assess how much one unit of methadone dose is by
finding the sample standard deviation:

```
summarize dose
```

```
 Variable |      Obs        Mean     Std. Dev.       Min          Max
 ---------+------------------------------------------------------------
     dose |      238     60.39916     14.45013        20          110
```

```
            failure _d:  status
    analysis time _t:  time

Iteration 0:    log likelihood = -614.68365
Iteration 1:    log likelihood = -597.73516
Iteration 2:    log likelihood =  -597.714
Refining estimates:
Iteration 0:    log likelihood =  -597.714

Stratified Cox regr. -- Breslow method for ties

No. of subjects =           238            Number of obs   =        238
No. of failures =           150
Time at risk    =         95812
                                           LR chi2(2)      =      33.94
Log likelihood  =      -597.714            Prob > chi2     =     0.0000

-------------------------------------------------------------------------
   _t |
   _d | Haz. Ratio   Std. Err.      z    P>|z|     [95% Conf. Interval]
------+------------------------------------------------------------------
 dose |  .9654655    .0062418   -5.436   0.000     .953309     .977777
prison|  1.475192    .2491827    2.302   0.021    1.059418    2.054138
-------------------------------------------------------------------------
                                                  Stratified by clinic
```

Display 11.1

indicating that a unit is not much at all; subjects often differ from each other by 10 to 15 units. To find the hazard ratio of two subjects differing by one standard deviation, we need to raise the hazard ratio to the power of one standard deviation, giving $0.9654655^{14.45013} = 0.60179167$. We can obtain the same result (with greater precision) using the stored macros _b[dose] for the log hazard ratio and r(Var) for the variance,

```
disp exp(_b[dose]*sqrt(r(Var)))
```

$$\boxed{.60178874}$$

In the above calculation, we simply rescaled the regression coefficient before taking the exponential. To obtain this hazard ratio in the Cox regression, standardize dose to have unit standard deviation. In the command below we also standardize to mean zero although this will make no difference to the estimated coefficients (except the constant):

```
egen zdose=std(dose)
```

we repeat the Cox regresion with the option bases(s) which results in the baseline survival function $S_0(t)$ being estimated and stored in s:

```
stcox zdose prison, strata(clinic) bases(s)
```

```
        failure _d:  status
   analysis time _t:  time

Iteration 0:   log likelihood = -614.68365
Iteration 1:   log likelihood = -597.73516
Iteration 2:   log likelihood =   -597.714
Refining estimates:
Iteration 0:   log likelihood =   -597.714


Stratified Cox regr. -- Breslow method for ties

No. of subjects =          238              Number of obs   =          238
No. of failures =          150
Time at risk    =        95812
                                            LR chi2(2)      =        33.94
Log likelihood  =     -597.714              Prob > chi2     =       0.0000


-------------------------------------------------------------------------------
  _t |
  _d | Haz. Ratio   Std. Err.      z    P>|z|     [95% Conf. Interval]
-----+-------------------------------------------------------------------------
zdose | .6017887    .0562195   -5.436   0.000     .5010998    .7227097
prison | 1.475192    .2491827    2.302   0.021    1.059418    2.054138
-------------------------------------------------------------------------------
                                                     Stratified by clinic
```

The coefficient of **zdose** is identical to that calculated previously and may now be interpreted as indicating a decrease of the hazard by 40% when the methadone dose increases by one standard deviation.

One question to investigate is whether the clinics need to be treated as strata or whether the hazard functions are in fact proportional. We may do this by plotting the log cumulative hazard function for the two clinics and visually assessing whether these curves are approximately parallel. The variables required for the graph are obtained as follows:

```
gen lh1=log(-log(s)) if clinic==1
gen lh2=log(-log(s)) if clinic==2
```

To find out where the two curves end so that we can label them at those point, we sort the data by time within clinic and list the last observation within clinic

```
sort clinic time
by clinic: list clinic status time lh1 lh2 if _n==_N
```

```
       -> clinic=        1
              clinic      status       time         lh1          lh2
       163.        1           0        905    1.477235            .

       -> clinic=        2
              clinic      status       time         lh1          lh2
       238.        2           0       1076           .    -.4802778
```

showing that the curves end at times 905 and 1075, respectively, with corresponding values on the vertical axis of 1.48 and -0.48. In order to plot labels "clin1" and "clin2" near these points, we can define new variables pos1 and pos2 equal to the y-positions for the two labels when time is equal to the required positions on the time-axis and equal to missing for all other values of time. We also need two string variables, lab1 and lab2 to hold the corresponding labels, to be applied at the points in pos1 and pos2 using the s() option.

```
gen pos1=1.48+0.2 if time==905
gen str5 lab1="clin1"
gen pos2=-0.48+0.2 if time==1076
gen str5 lab2="clin2"
```

The graph is produced using

```
graph lh1 lh2 pos1 pos2 time, sort s(ii[lab1][lab2]) c(JJ..) /*
    */ xlabel ylabel pen(6311) psize(130) t1(" ") gap(3)    /*
    */ l1("Log of cumulative hazard")
```

Here, the option connect(J), abbreviated c(J), causes points to be connected using a step function and pen(6311) was used to control the colors (and thicknesses) of the lines and plotting symbols, mainly to have the labels appear in the same color (pen=1) as the axes and titles. The resulting graph is shown in Figure 11.1. In these curves, the increment at each failure represents the estimated log of the hazards at that time. Clearly, the curves are not parallel and we will therefore continue treating the clinics as strata. Note that a quicker way of producing a similar graph would be to use the stphplot command as follows:

```
stphplot, strata(clinic) xlabel ylabel adjust(zdose prison) /*
    */ zero gap(3) psize(130)
```

Here the adjust() option specifies the covariates to be used in the Cox regression, and the zero option specifies that these covariates are to be evaluated at zero. As a result, minus the log of the baseline survival functions (stratified by clinics) are plotted agains the log of the survival time, see Figure 11.2.

11.3.1 Model presentation and diagnostics

Assuming the variables prison and zdose satisfy the proportional hazards assumption (see Section 11.3.2), we now discuss how to present the model.

A good graph for presenting the results of a Cox regression is a graph of the survival curves fitted for groups of subjects when the continuous variables take on their mean values. Such a graph may be produced by first generating

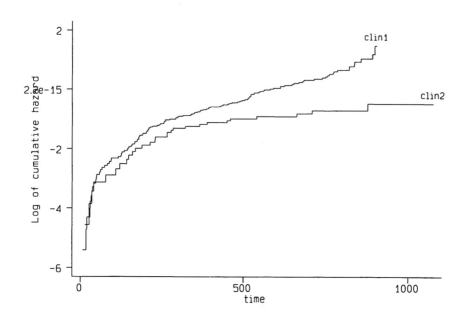

Figure 11.1 *Log of minus the log of the survival functions for the two clinics estimated by stratified Cox regression.*

variables surv1 and surv2 to hold the fitted baseline survival curves for the two groups and then applying equation (11.8) to obtain the survival curves for particular covariate values:

```
stcox zdose prison, strata(clinic) bases(s_strat)
egen mdose=mean(zdose), by(clinic)
gen surv1=s_strat^exp(_b[zdose]*mdose) if prison==0
gen surv2=s_strat^exp(_b[zdose]*mdose+_b[prison]) if prison==1
```

Note that the survival functions represent the expected survival function for a subject having the clinic-specific mean dose. We now transform time to time in years and plot the survival curves separately for each clinic:

```
gen tt=time/365.25
label variable tt "time in years"
graph surv1 surv2 tt, sort c(JJ) s(..) by(clinic) t1(" ") /*
    */ xlabel ylabel(0.0,0.2,0.4,0.6,0.8,1.0)
```

resulting in two graphs displayed side by side at the top of the graphics window separated by a wide gap from a title at the bottom of the window. A better

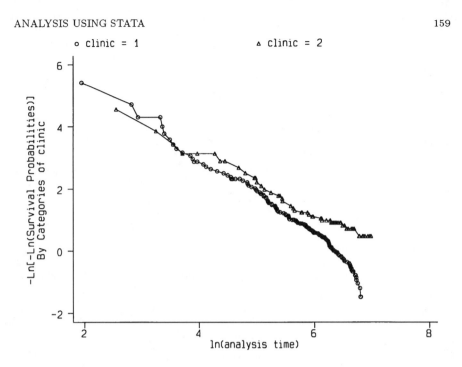

Figure 11.2 *Minus the log of minus the log of the survival functions for the two clinics versus log survival time, estimated by stratified Cox regression.*

looking figure can be produced by displaying separate graphs for the clinics side by side where, the separate graphs are created using

```
graph surv1 surv2 tt if clinic==1 , sort c(JJ) s(..) gap(3) /*
  */ xlabel(0,1,2,3) ylabel(0.0,0.2,0.4,0.6,0.8,1.0)     /*
  */ t1("clinic 1") l1("fraction remaining")
```

and similarly for clinic 2. The result is shown in Figure 11.3.

According to Caplehorn and Bell (1991), the more rapid decline in the proportion remaining in clinic 1 compared with clinic 2 may be due to the policy of clinic 1 to attempt to limit the duration of maintenance to two years.

It is probably a good idea to produce some residual plots, for example a graph of the deviance residuals against the linear predictor. In order for **predict** to be able to compute the deviance residuals, we must first store the martingale residuals (see for example Collett, 1994) using **stcox** with the **mgale()** option:

```
stcox zdose prison, strata(clinic) mgale(mart)
predict devr, deviance
predict xb, xb
graph devr xb, xlabel ylabel s([id]) gap(3) /*
```

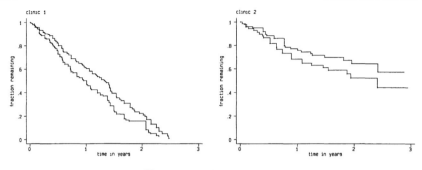

Figure 11.3 *Survival curves.*

```
*/ 11("Deviance Residuals")
```

with the result shown in Figure 11.4. There appear to be no serious outliers.

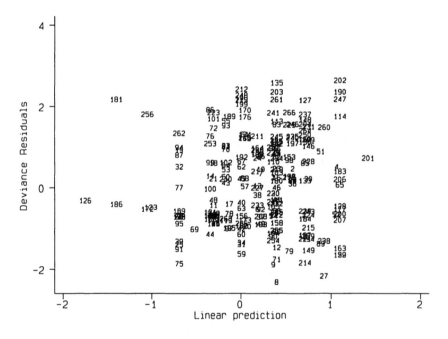

Figure 11.4 *Deviance residuals for survival analysis.*

Another type of residual is a score residual, defined as the first derivative of the partial log likelihood function with respect to an explanatory variable. The score residual is large in absolute value if a case's explanatory variable differs substantially from the explanatory variables of subjects whose estimated risk of failure is large at the case's time of failure or censoring. Since our model has two explanatory variables, we can compute the score residuals for `zdose` and `prison` and store them in `score1` and `score2` using the command

```
drop s_strat
stcox zdose prison, strata(clinic) bases(s_strat) esr(score*)
```

These residuals can be plotted against survival time using

```
label variable score1 "score residuals for zdose"
graph score1 tt, s([id]) gap(3) xlabel ylabel
```

and similarly for `score2`. The resulting graphs are shown in Figures 11.5 and 11.6. Subject 89 has a low value of `zdose` (-1.75) compared with other subjects at risk of failure at such a late time. Subjects 8, 27, 12, and 71 leave relatively late considering that they have a police record whereas other remainers at their time of leaving (or censoring) tend to have no police record.

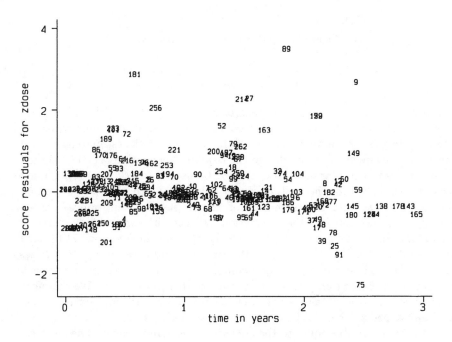

Figure 11.5 *Score residuals for zdose.*

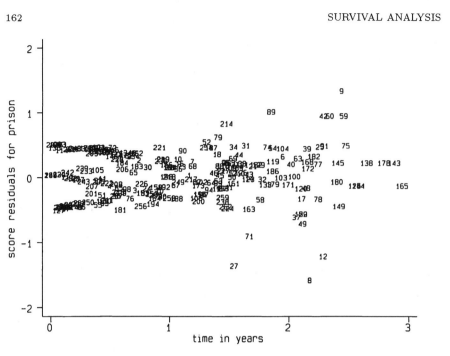

Figure 11.6 *Score residuals for prison.*

11.3.2   Cox's regression with time-varying covariates

In order to determine whether the hazard functions for those with and without
a prison history are proportional, we could split the data into four strata by
clinic and prison. However, as the strata get smaller, the estimated survival
functions become less precise (due to the risk sets in equation (11.6) becoming
smaller). Also, a similar method could not be used to check the proportional
hazard assumption for the continuous variable zdose. Another way of testing
the proportional hazards assumption of zdose, say, is to introduce a time-
varying covariate equal to the interaction between the survival time variable
and zdose. In order to estimate this model, the terms in equation (11.6) need
to be evaluated using the values of the time-varying covariates *at the times
of the failure.* These values are not available in the present dataset since each
subject is represented only once, at the time of their own failure. We therefore
need to expand the dataset so that each subject's record appears (at least)
as many times as that subject contributes to a risk set in equation (11.6),
with the time variable equal to the corresponding failure times. The simplest
way to achieve this for discrete survival times is to represent each subject by t_i
observations (where t_i is subject i's survival time) with the variable time equal
to $1,2,\cdots,t_i$. The invented survival times are times beyond which the subject

survives, so the status variable `fail`, is set to zero for all invented times and equal to `status` for the original survival time. The following code is adapted from the Stata FAQ by W. Gould (1999) (to look at the FAQ, do a search with keyword equal to "time-varying," click into url of the appropriate FAQ entry).

```
expand time
sort id
quietly by id: gen t = _n
gen fail = 0
quietly by id: replace fail = status if _n==_N
```

The same may be achieved with a more concise syntax:

```
stset time, failure(status) id(id)
quietly summarize _t, meanonly
stsplit cat, at(0(1)`r(max)´)
```

A problem with applying this code to our data is that our survival times are measured in days with a median survival of 504 days. Unless we are willing to round time, for example to integer number of months, we would have to replicate the data by a factor of about 500! Therefore, a more feasible method is described in several stages.

Before we begin, we must make sure that we have enough memory allocated to Stata to hold the increased dataset. The memory may be set to 3 mega-bytes using

```
clear
set memory 3m
```

(the data have to be read in again after this). If `set memory` does not work, add the `/k3000` option to the command for running Stata, e.g.,

```
C:\Stata\wstata.exe /k3000
```

in the **Target** field of the **Shortcut** tab accessed by right-mouse clicking into the Stata icon and selecting **Properties**. We will also need to define matrices and in order to ensure that we do not exceed the maximum matrix size, we increase the limit to 300 using `set matsize`:

```
set matsize 300
```

The first step is to determine the sets of unique failure times for the two clinics and store these in matrices `t1` and `t2` using the command `mkmat`:

```
sort clinic status time
gen unft=0 /* indicator for unique failure times */
quietly by clinic status time: replace unft=1 if _n==1&status==
```

```
sort time
mkmat time if (unft==1)&clinic==1,matrix(t1)
sort time
mkmat time if (unft==1)&clinic==2,matrix(t2)
```

We can look at the contents of t2 using

```
matrix list t2
```

with the result shown in Display 11.2. Note that t2 is a matrix although it

```
t2[27,1]
          time
  r1       13
  r2       26
  r3       35
  r4       41
  r5       79
  r6      109
  r7      122
  r8      143
  r9      149
 r10      161
 r11      170
 r12      190
 r13      216
 r14      231
 r15      232
 r16      268
 r17      280
 r18      286
 r19      322
 r20      366
 r21      389
 r22      450
 r23      460
 r24      540
 r25      661
 r26      708
 r27      878
```

Display 11.2

only holds a one-dimensional array of numbers. This is because there are no vectors in Stata. The matrix is $r \times 1$ where r can be found using the function rowsof(t2). Elements of the matrix may be accessed using the expression t2[3,1] for the third unique failure time in clinic2.

Next, determine num, the number of replicates required of each subject's record. Initially, num is set equal to the number of unique survival times in each subject's clinic, i.e., the number of rows of the corresponding matrix:

```
local nt1=rowsof(t1)
```

```
local nt2=rowsof(t2)
gen num=cond(clinic==1,`nt1´,`nt2´)
```

However, the number of required replicates is just the number of unique survival times *which are less than* the subject's own survival time, time because the subject ceases to contribute to future risk sets once it has failed (or has been censored). Therefore, reduce num until t1[num,1] is less than time for the first time, implying that t1[num+1,1] is greater than time:

```
local i=1
while `i´<=`nt1´
    qui replace num=num-1                              /*
        */ if clinic==1&time<=t1[`nt1´-`i´+1,1]
    local i=`i´+1


local i=1
while `i´<=`nt2´
    qui replace num=num-1                              /*
        */ if clinic==2&time<=t2[`nt2´-`i´+1,1]
    local i=`i´+1
```

We need one extra record for each subject to hold their own survival time:

```
replace num=num+1
```

We are now ready to expand the dataset and fill in the unique survival times for the first num-1 records of each subject, followed by the subject's own survival time in the last record. All except the last values of the new variable fail (for failure status) will be 0 (for censored) and the last value will be equal to the subject's status at the time of their own failure or censoring:

```
compress
disp _N
expand num
disp _N

sort id time
quietly by id: gen t=cond(clinic==1,t1[_n,1],t2[_n,1])
sort id t
quietly by id: replace t=time if _n==_N

gen fail=0 /* indicator for failure */
sort id t
quietly by id: replace fail=status if _n==_N
```

The command `compress` was used before `expand` to ensure that the variables are stored as compactly as possible before another 10990 observations were created. The data are now in a suitable form to perfom Cox regression with time-varying covariates. We first verify that the data are equivalent to the original form by repeating the simple Cox regression

```
stset t, failure(fail) id(id)
stcox zdose prison, strata(clinic)
```

which gives the same result as before. We can now generate a time-varying covariate for zdose as follows:

```
gen tdose=zdose*(t-504)/365.25
```

where we have subtracted the median survival time so that the effect of the explanatory variable zdose can be interpreted as the effect of zdose at the median survival time. We have divided by 365.25 to see by how much the effect of zdose changes between intervals of one year. We now fit the Cox regression, allowing the effect of zdose to vary with time:

```
stcox zdose tdose prison, strata(clinic)
```

In the output shown in Display 11.3, the estimated increase in the hazard ratio for zdose is 15% per year. This small effect is not significant at the 5% level which is confirmed by carrying out the likelihood ratio test as follows:

```
lrtest, saving(0)
quietly stcox zdose prison, strata(clinic)
lrtest
```

Cox: likelihood-ratio test	chi2(1) =	0.85
	Prob > chi2 =	0.3579

giving a very similar p-value as before and confirming that there is no evidence that the effect of dose on the hazard varies with time. A similar test can be carried out for prison (see Exercise 5).

We can now return back to the original data, either by reading them again or by dropping all the newly created observations as follows:

```
sort id t
quietly by id: drop if _n<_N
stset t, failure(fail)
stcox zdose prison, strata(clinic) bases(s_strat)
```

11.4 Exercises

1. In the original analysis of this data, Caplehorn and Bell (1991) judged that the hazards were approximately proportional for the first 450 days (see

```
           failure _d:  fail
    analysis time _t:  t
                id:  id

Iteration 0:    log likelihood = -614.68365
Iteration 1:    log likelihood = -597.32655
Iteration 2:    log likelihood = -597.29131
Iteration 3:    log likelihood = -597.29131
Refining estimates:
Iteration 0:    log likelihood = -597.29131


Stratified Cox regr. -- Breslow method for ties

No. of subjects =          238          Number of obs   =      11228
No. of failures =          150
Time at risk    =        95812
                                         LR chi2(3)      =      34.78
Log likelihood  =   -597.29131          Prob > chi2     =     0.0000

------------------------------------------------------------------------------
     _t |
     _d |  Haz. Ratio  Std. Err.      z     P>|z|    [95% Conf. Interval]
--------+---------------------------------------------------------------------
  zdose |   .6442974   .0767535    -3.690   0.000     .5101348    .8137442
  tdose |   1.147853   .1720104     0.920   0.357     .8557175   1.539722
 prison |   1.481193   .249978      2.328   0.020     1.064036   2.061899
------------------------------------------------------------------------------
                                                     Stratified by clinic
```

Display 11.3

Figure 11.1). They therefore analyzed the data for this time period using
clinic as a covariate instead of stratifying by clinic. Repeat this analysis,
using prison and dose as further covariates.

2. Following Caplehorn and Bell (1991), repeat the above analysis treating
   dose as a categorical variable with three levels (< 60, $60 - 79$, ≥ 80) and
   plot the predicted survival curves for the three dose categories when prison
   and clinic take on one of their values.

3. Test for an interaction between clinic and the methadone dose using both
   continuous and categorical scales for dose.

4. Create a "do-file" containing the commands given above to expand the data
   for Cox regression with time-varying covariates.

5. Read and expand the data (using the "do-file") and check the proportional
   hazards assumption for prison using the same method we used for dose.

CHAPTER 12

Principal Components Analysis: Hearing Measurement using an Audiometer

12.1 Description of data

The data in Table 12.1 are adapted from those given in Jackson (1991), and relate to hearing measurement with an instrument called an audiometer. An individual is exposed to a signal of a given frequency with an increasing intensity until the signal is perceived. The lowest intensity at which the signal is perceived is a measure of hearing loss, calibrated in units referred to as *decibel loss* in comparison to a reference standard for that particular instrument. Observations are obtained one ear at a time for a number of frequencies. In this example, the frequencies used were 500 Hz, 1000 Hz, 2000 Hz, and 4000 Hz. The limits of the instrument are -10 to 99 decibels. (A negative value does not imply better than average hearing; the audiometer had a calibration 'zero' and these observations are in relation to that.)

Table 12.1: Data in *hear.dat* (Taken from Jackson (1991) with permission of his publisher, John Wiley & Sons).

id	l500	l1000	l2000	l4000	r500	r1000	r2000	r4000
1	0	5	10	15	0	5	5	15
2	-5	0	-10	0	0	5	5	15
3	-5	0	15	15	0	0	5	15
4	-5	0	-10	-10	-10	-5	-10	10
5	-5	-5	-10	10	0	-10	-10	50
6	5	5	5	-10	0	5	0	20
7	0	0	0	20	5	5	5	10
8	-10	-10	-10	-5	-10	-5	0	5
9	0	0	0	40	0	0	-10	10
10	-5	-5	-10	20	-10	-5	-10	15
11	-10	-5	-5	5	5	0	-10	5
12	5	5	10	25	-5	-5	5	15
13	0	0	-10	15	-10	-10	-10	10
14	5	15	5	60	5	5	0	50
15	5	0	5	15	5	-5	0	25
16	-5	-5	5	30	5	5	5	25

Table 12.1: Data in *hear.dat* (continued)

17	0	-10	0	20	0	-10	-10	25
18	5	0	0	50	10	10	5	65
19	-10	0	0	15	-10	-5	5	15
20	-10	-10	-5	0	-10	-5	-5	5
21	-5	-5	-5	35	-5	-5	-10	20
22	5	15	5	20	5	5	5	25
23	-10	-10	-10	25	-5	-10	-10	25
24	-10	0	5	15	-10	-5	5	20
25	0	0	0	20	-5	-5	10	30
26	-10	-5	0	15	0	0	0	10
27	0	0	5	50	5	0	5	40
28	-5	-5	-5	55	-5	5	10	70
29	0	15	0	20	10	-5	0	10
30	-10	-5	0	15	-5	0	10	20
31	-10	-10	5	10	0	0	20	10
32	-5	5	10	25	-5	0	5	10
33	0	5	0	10	-10	0	0	0
34	-10	-10	-10	45	-10	-10	5	45
35	-5	10	20	45	-5	10	35	60
36	-5	-5	-5	30	-5	0	10	40
37	-10	-5	-5	45	-10	-5	-5	50
38	5	5	5	25	-5	-5	5	40
39	-10	-10	-10	0	-10	-10	-10	5
40	10	20	15	10	25	20	10	20
41	-10	-10	-10	20	-10	-10	0	5
42	5	5	-5	40	5	10	0	45
43	-10	0	10	20	-10	0	15	10
44	-10	-10	10	10	-10	-10	5	0
45	-5	-5	-10	35	-5	0	-10	55
46	5	5	10	25	10	5	5	20
47	5	0	10	70	-5	5	15	40
48	5	10	0	15	5	10	0	30
49	-5	-5	5	-10	-10	-5	0	20
50	-5	0	10	55	-10	0	5	50
51	-10	-10	-10	5	-10	-10	-5	0
52	5	10	20	25	0	5	15	0
53	-10	-10	50	25	-10	-10	-10	40
54	5	10	0	-10	0	5	-5	15
55	15	20	10	60	20	20	0	25
56	-10	-10	-10	5	-10	-10	-5	-10
57	-5	-5	-10	30	0	-5	-10	15
58	-5	-5	0	5	-5	-5	0	10
59	-5	5	5	40	0	0	0	10
60	5	10	30	20	5	5	20	60
61	5	5	0	10	-5	5	0	10

Table 12.1: Data in *hear.dat* (continued)

62	0	5	10	35	0	0	5	20
63	-10	-10	-10	0	-5	0	-5	0
64	-10	-5	-5	20	-10	-10	-5	5
65	5	10	0	25	5	5	0	15
66	-10	0	5	60	-10	-5	0	65
67	5	10	40	55	0	5	30	40
68	-5	-10	-10	20	-5	-10	-10	15
69	-5	-5	-5	20	-5	0	0	0
70	-5	-5	-5	5	-5	0	0	5
71	0	10	40	60	-5	0	25	50
72	-5	-5	-5	-5	-5	-5	-5	5
73	0	5	45	50	0	10	15	50
74	-5	-5	10	25	-10	-5	25	60
75	0	-10	0	60	15	0	5	50
76	-5	0	10	35	-10	0	0	15
77	5	0	0	15	0	5	5	25
78	15	15	5	35	10	15	-5	0
79	-10	-10	-10	5	-5	-5	-5	5
80	-10	-10	-5	15	-10	-10	-5	5
81	0	-5	5	35	-5	-5	5	15
82	-5	-5	-5	10	-5	-5	-5	5
83	-5	-5	-10	-10	0	-5	-10	0
84	5	10	10	20	-5	0	0	10
85	-10	-10	-10	5	-10	-5	-10	20
86	5	5	10	0	0	5	5	5
87	-10	0	-5	-10	-10	0	0	-10
88	-10	-10	10	15	0	0	5	15
89	-5	0	10	25	-5	0	5	10
90	5	0	-10	-10	10	0	0	0
91	0	0	5	15	5	0	0	5
92	-5	0	-5	0	-5	-5	-10	0
93	-5	5	-10	45	-5	0	-5	25
94	-10	-5	0	10	-10	5	-10	10
95	-10	-5	0	5	-10	-5	-5	5
96	5	0	5	0	5	0	5	15
97	-10	-10	5	40	-10	-5	-10	5
98	10	10	15	55	0	0	5	75
99	-5	5	5	20	-5	5	5	40
100	-5	-5	-10	-10	-5	0	15	10

12.2 Principal component analysis

Principal component analysis is one of the oldest but still most widely used techniques of multivariate analysis. Originally introduced by Pearson (1901)

and independently by Hotelling (1933), the basic idea of the method is to
try to describe the variation of the variables in a set of multivariate data as
parsimoniously as possible using a set of derived uncorrelated variables, each of
which is a particular linear combination of those in the original data. In other
words, principal components analysis is a transformation from the observed
variables, $x_1, \cdots, x_p$ to new variables $y_1, \cdots, y_p$ where

$$
\begin{aligned}
y_1 &= a_{11}x_1 + a_{12}x_2 + \cdots + a_{ip}x_p \\
y_2 &= a_{21}x_1 + a_{22}x_2 + \cdots + a_{2p}x_p \\
\vdots &= \vdots + \vdots + \vdots + \vdots \\
y_p &= a_{p1}x_1 + a_{p2}x_2 + \cdots + a_{pp}x_p.
\end{aligned}
\tag{12.1}
$$

The new variables are derived in decreasing order of importance so that the
first principal component y_1 accounts for as much of the variation of the original
variables as possible. The second principal component y_2 accounts for as much
of the remaining variation as possible *conditional* on being uncorrelated with
y_1 and so on. The usual objective of this type of analysis is to assess whether
the first few components account for a substantial proportion of the variation
in the data. If they do, they can be used to summarize the data with little
loss of information. This may be useful for obtaining graphical displays of the
multivariate data or for simplifying subsequent analysis.

The coefficients defining the principal components are obtained from the
eigenvalues of either the covariance or correlation matrix of the original variables (giving different results). The variances of the derived variables are given
by the eigenvalues of the corresponding matrix. A detailed account of principal
components analysis is given in Everitt and Dunn (1991).

12.3 Analysis using Stata

The data can be read in from an ASCII file *hear.dat* as follows:

```
infile id 1500 11000 12000 14000 r500 r1000 r2000 r4000 /*
     */ using hear.dat
summarize
```

Variable	Obs	Mean	Std. Dev.	Min	Max
id	100	50.5	29.01149	1	100
1500	100	-2.8	6.408643	-10	15
11000	100	-.5	7.571211	-10	20
12000	100	2.45	11.94463	-10	50
14000	100	21.35	19.61569	-10	70
r500	100	-2.6	7.123726	-10	25
r1000	100	-.7	6.396811	-10	20
r2000	100	1.55	9.257675	-10	35
r4000	100	20.95	19.43254	-10	75

Before undertaking a principal components analysis, some graphical exploration of the data may be useful. A scatter-plot matrix, for example, with points labeled with a subject's identification number can be obtained using

```
graph l500-r4000, matrix half symbol([id]) ps(150)
```

The resulting diagram is shown in Figure 12.1. The diagram looks a little 'odd' due to the largely 'discrete' nature of the observations. Some of the individual scatterplots suggest that some of the observations might perhaps be regarded as outliers; for example, individual 53 in the plot involving l2000, r2000. This subject has a score of 50 at this frequency in the left ear, but a score of -10 in the right ear. It might be appropriate to remove this subject's observations before further analysis, but we shall not do this and will continue to use the data from all 100 individuals.

As mentioned in the previous section, principal components may be extracted from either the covariance matrix or the correlation matrix of the original variables. A choice needs to be made since there is not necessarily any simple relationship between the results in each case. The summary table shows that the variances of the observations at the highest frequencies are approximately nine times those at the lower frequencies; consequently, a principal components analysis using the covariance matrix would be dominated by the 4000 Hz frequency. But this frequency is not more clinically important than the others, and so, in this case, it seems more reasonable to use the correlation matrix as the basis of the principal components analysis.

To find the correlation matrix of the data requires the following instruction:

```
correlate l500-r4000
```

and the result is given in Display 12.1. Note that the highest correlations occur between adjacent frequencies on the same ear and between corresponding frequencies on different ears.

To obtain the principal components of this correlation matrix requires the use of the factor procedure:

```
factor l500-r4000, pc
```

which give the results shown in Display 12.2. (The pc option was used to obtain principal components rather than a factor analysis solution.)

An informal rule for choosing the number of components to represent a set of correlations is to use only those components with eigenvalues greater than one. Here, this would lead to retaining only the first two components. Another informal indicator of the appropriate number of components is the *scree-plot*, a plot of the eigenvalues. A scree-plot may be obtained using

```
greigen, gap(3)
```

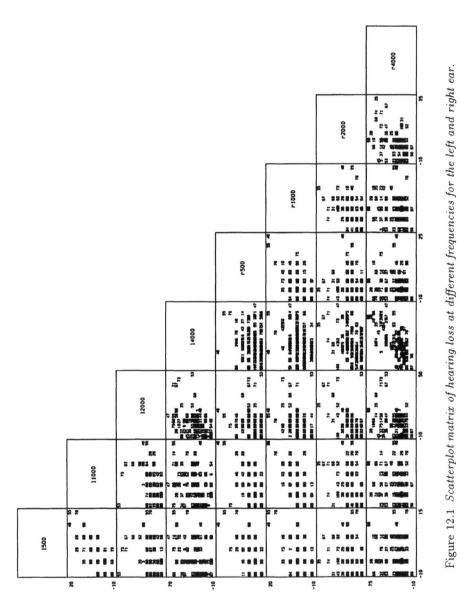

Figure 12.1 Scatterplot matrix of hearing loss at different frequencies for the left and right ear.

```
(obs=100)

           |    l500    l1000    l2000    l4000      r500     r1000     r2000
-----------+-----------------------------------------------------------------
     l500|  1.0000
    l1000|  0.7775   1.0000
    l2000|  0.3247   0.4437   1.0000
    l4000|  0.2554   0.2749   0.3964   1.0000
     r500|  0.6963   0.5515   0.1795   0.1790   1.0000
    r1000|  0.6416   0.7070   0.3532   0.2632   0.6634   1.0000
    r2000|  0.2399   0.3606   0.5910   0.3193   0.1575   0.4151   1.0000
    r4000|  0.2264   0.2109   0.3598   0.6783   0.1421   0.2248   0.4044

           |   r4000
-----------+----------
    r4000|  1.0000
```

Display 12.1

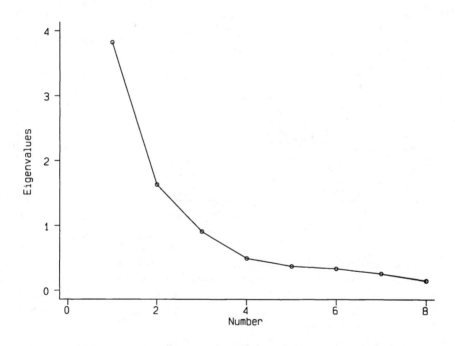

Figure 12.2 *Scree-plot.*

with the result shown in Figure 12.2. An 'elbow' in the scree-plot indicates
the number of eigenvalues to choose. From Figure 9.2, this would again appear
to be two. The first two components account for 68% of the variance in the
data.

```
(obs=100)

          (principal components; 8 components retained)
Component    Eigenvalue    Difference    Proportion    Cumulative
-----------------------------------------------------------------
    1          3.82375       2.18915        0.4780        0.4780
    2          1.63459       0.72555        0.2043        0.6823
    3          0.90904       0.40953        0.1136        0.7959
    4          0.49951       0.12208        0.0624        0.8584
    5          0.37743       0.03833        0.0472        0.9055
    6          0.33910       0.07809        0.0424        0.9479
    7          0.26101       0.10545        0.0326        0.9806
    8          0.15556          .           0.0194        1.0000


          Eigenvectors
Variable |    1          2          3          4          5          6
---------+-------------------------------------------------------------------
   1500 |  0.40915    -0.31257    0.13593   -0.27217   -0.16650    0.41679
  11000 |  0.42415    -0.23011   -0.09332   -0.35284   -0.49977   -0.08474
  12000 |  0.32707     0.30065   -0.47772   -0.48723    0.50331    0.04038
  14000 |  0.28495     0.44875    0.47110   -0.17955    0.09901   -0.51286
   r500 |  0.35112    -0.38744    0.23944    0.30453    0.62830    0.17764
  r1000 |  0.41602    -0.23673   -0.05684    0.36453   -0.08611   -0.54457
  r2000 |  0.30896     0.32280   -0.53841    0.51686   -0.16229    0.12553
  r4000 |  0.26964     0.49723    0.41499    0.19757   -0.17570    0.45889


          Eigenvectors
Variable |    7          8
---------+---------------------
   1500 |  0.28281    -0.60077
  11000 | -0.02919     0.61330
  12000 | -0.27925    -0.06396
  14000 |  0.43536    -0.02978
   r500 |  0.12745     0.36603
  r1000 | -0.46180    -0.34285
  r2000 |  0.44761     0.02927
  r4000 | -0.47094     0.07469
```

Display 12.2

The elements of the eigenvectors defining the principal components are scaled so that their sums of squares are unity. A more useful scaling is often obtained from multiplying the elements by the square root of the corresponding eigenvalue, in which case the coefficients represent correlations between an observed variable and a component.

Examining the eigenvectors defining the first two principal components, we see that the first accounting for 48% of the variance has coefficients that are all positive and all approximately the same size. This principal component essentially represents the overall hearing loss of a subject and implies that individuals suffering hearing loss at certain frequencies will be more likely to suffer this loss at other frequencies as well. The second component, accounting for 20% of the variance, contrasts high frequencies (2000 Hz and 4000 Hz) and low frequencies (500 Hz and 1000 Hz). It is well known in the case of normal hearing that hearing loss as a function of age is first noticeable in the higher frequencies.

Scores for each individual on the first two principal components might be used as a convenient way of summarizing the original eight-dimensional data. Such scores are obtained by applying the elements of the corresponding eigenvector to the standardized values of the original observations for an individual. The necessary calculations can be carried out with the **score** procedure:

```
score pc1 pc2
```

```
            (based on unrotated principal components)
            (6 scorings not used)

            Scoring Coefficients
Variable |     1          2
---------+------------------------
   1500  |  0.40915    -0.31257
  11000  |  0.42415    -0.23011
  12000  |  0.32707     0.30065
  14000  |  0.28495     0.44875
   r500  |  0.35112    -0.38744
  r1000  |  0.41602    -0.23673
  r2000  |  0.30896     0.32280
  r4000  |  0.26964     0.49723
```

The new variables *pc1* and *pc2* contain the scores for the first two principal components and the output lists the coefficients used to form these scores. For principal component analysis, these coefficients are just the elements of the eigenvectors in Display 12.2. The principal component scores can be used to produce a useful graphical display of the data in a single scatter-plot, which may then be used to search for structure or patterns in the data, particularly the presence of clusters of observations (see Everitt (1993)). Note that the distances between observations in this graph approximate the Euclidean distances between the (standardized) variables, i.e., the graph is a *multidimensional scaling* solution. In fact, the graph is the classical scaling (or principal coordinate) scaling solution to the Euclidean distances (see Everitt and Dunn (1991) or Everitt and Rabe-Hesketh (1997)).

The principal component plot is obtained using

```
graph pc2 pc1, symbol([id]) ps(150) gap(3) xlab ylab
```

The resulting diagram is shown in Figure 12.3. Here, the variablility in differential hearing loss for high versus low frequencies (pc2) is greater among subjects with higher overall hearing loss, as would be expected. It would be interesting to investigate the relationship between the principal components and other variables related to hearing loss such as age (see Exercise 6).

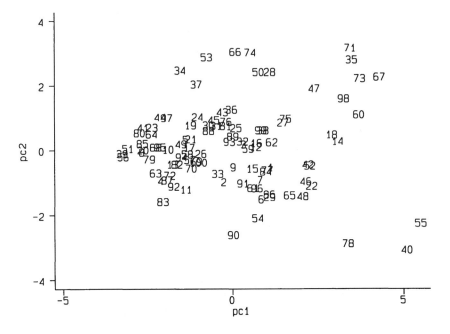

Figure 12.3 *Principal component plot.*

12.4 Exercises

1. Rerun the principal component analysis described in this chapter using the covariance matrix of the observations. Compare the results with those based on the correlation matrix.

2. Interpret components 3 through 8 in the principal components analysis based on the correlation matrix.

3. Create a scatterplot matrix of the first five principal components scores.

4. Investigate other methods of factor analysis available in `factor` applied to the hearing data.

5. Apply principal component analysis to the air pollution data analyzed in Chapter 3, exluding the variable so2, and plot the first two principal components (i.e., the 2-d classical scaling solution for Euclidean distances between standardized variables).

6. Regress so2 on the first two principal components and add a line corresponding to this regression (the direction of steepest increase in so2 predicted by the regression plane) into the multidimenional scaling solution.

CHAPTER 13

Maximum Likelihood Estimation: Age of Onset of Schizophrenia

13.1 Description of data

In Table 13.1, the onset ages of schizophrenia in 99 women (determined as age on first admission) are given. These data will be used in order to investigate whether there is any evidence for the subtype model of schizophrenia (see Lewine, 1981). According to this model, there are two types of schizophrenia. One is characterized by early onset, typical symptoms and poor premorbid competence and the other by late onset, atypical symptoms and good premorbid competence. We will investigate this question by fitting a mixture of two normal distributions to the ages.

13.2 Finite mixture distributions

The most common type of finite mixture distribution encountered in practice is a mixture of univariate Gaussian distributions of the form

$$f(y_i; \mathbf{p}, \boldsymbol{\mu}, \boldsymbol{\sigma}) = p_1 g(y_i; \mu_1, \sigma_1) + p_2 g(y_i; \mu_2, \sigma_2) + \cdots + p_k g(y_i; \mu_k, \sigma_k) \quad (13.1)$$

where $g(y; \mu, \sigma)$ is the Gaussian density with mean μ and standard deviation σ,

$$g(y; \mu, \sigma) = \frac{1}{\sigma\sqrt{2\pi}} \exp\left\{ -\frac{1}{2}\left(\frac{y - \mu}{\sigma}\right)^2 \right\} \quad (13.2)$$

and $p_1, \cdots, p_k$, are the mixing probabilities.

The parameters $p_1, \cdots, p_k$, $\mu_1, \cdots, \mu_k$ and $\sigma_1, \cdots, \sigma_k$ are usually estimated by maximum likelihood. Standard errors may be obtained in the usual way from the observed information matrix (i.e., the inverse of the Hessian matrix, the matrix of second derivatives of the log-likelihood). Determining the number of components in the mixture, k, is more problematic. This question is not considered here.

For a short introduction to finite mixture modeling, see Everitt (1996); a more comprehensive account is given in McLachlan and Basford (1988).

Table 13.1: Onset age of schizophrenia in 99 women. The data are in *onset.dat*

Age of onset			
20	30	21	23
30	25	13	19
16	25	20	25
27	43	6	21
15	26	23	21
23	23	34	14
17	18	21	16
35	32	48	53
51	48	29	25
44	23	36	58
28	51	40	43
21	48	17	23
28	44	28	21
31	22	56	60
15	21	30	26
28	23	21	20
43	39	40	26
50	17	17	23
44	30	35	20
41	18	39	27
28	30	34	33
30	29	46	36
58	28	30	28
37	31	29	32
48	49	30	

13.3 Analysis using Stata

Stata has a set of commands, called ml, which can be used to maximize a user-specified log-likelihood using a modified Newton-Raphson algorithm. The algorithm is iterative. Starting with initial parameter estimates, the program evaluates the first and second derivatives of the log-likelihood in order to update the parameters in such a way that the log-likelihood is likely to increase. The updated parameters are then used to re-evalute the derivatives etc., until the program has converged to the maximum. The user needs to write a program which enables the ml program to evaluate the log-likelihood and its derivatives.

There are four alternative methods available with ml, d0, d1, d2, and lf. The method d0 does not require the user's program to evaluate any derivatives of the log-likelihood, d1 requires first derivatives only, and d2 requires both first and second derivatives. When d0 is chosen, first and second derivatives are determined numerically; this makes this alternative slower and less accurate than d1 or d2.

The simplest approach to use is lf which also does not require any derivatives to be programmed. Instead, the structure of most likelihood problems is used to increase both the speed and accuracy of the numerical estimation of first and second derivatives of the log-likelihood function. Whereas d0 to d2 can be used for any maximum likelihood problem, method lf may only be used if the likelihood satisfies the following two criteria:

1. The observations in the dataset are independent, i.e., the log-likelihood is the sum of the log-likelihood contributions of the observations.

2. The log-likelihood contributions have a *linear form*, i.e., they are (not necessarily linear) functions of linear predictors of the form $\eta_i = x_{1i}\beta_i + \cdots + x_{ki}\beta_k$.

The first restriction is usually met, an exception being clustered data. The second restriction is not as severe as it appears because there may be several linear predictors, as we shall see later.

In this chapter, we will give only a brief introduction to maximum likelihood estimation using Stata, restricting our examples to the lf method. We recommend the book on *Maximum Likelihood Estimation with Stata* by Gould and Sribney (1999) for a thorough treatment of the topic.

We will eventually fit a mixture of normals to the age of onset data, but will introduce the ml procedure by a series of more simple steps. To begin with, we will fit a normal distribution with standard deviation fixed at 1 to a set of simulated data. A (pseudo)-random sample from the normal distribution with mean 5 and variance 1 may be obtained using the instructions

```
clear
set obs 100
set seed 12345678
gen y = invnorm(uniform())+5
```

where the purpose of the set seed command is simply to ensure that the same data will be generated each time. We use **summarize** to confirm that the sample has a mean close to 5 and a standard deviation close to 1.

```
summarize y
```

Variable	Obs	Mean	Std. Dev.	Min	Max
y	100	5.002311	1.053095	2.112869	7.351898

First, define a program, mixing0, to evaluate the likelihood contributions when called from ml. The function must have two arguments; the first is the variable name where ml will look for the computed log-likelihood contributions;

the second is the variable name containing the 'current' value of the linear
predictor during the iterative maximization procedure:

```
capture program drop mixing0
program define mixing0
    version 6
    args lj xb

    tempname s
    scalar `s´  = 1
    tempvar f

    quietly gen double `f´ = /*
    */ exp(-.5*(($ML_y1-`xb´)/`s´)^2)/(sqrt(2*_pi)*`s´)
    quietly replace `lj´ = ln(`f´)
end
```

After giving names lj and xb to the arguments passed to mixing0 by ml,
mixing0 defines a temporary name stored in the local macro s and sets it equal
to the standard deviation (1) in the next command. A temporary variable with
its name stored in the local macro f is then defined to hold the normal density
for each observation. The reason for defining s and f is to make subsequent
commands easier to read. Using temporary names avoids any confusion with
variables that may exist in the dataset and causes the scalar and variable to
disappear when the program has finished running.

The normal density in equation (13.2) is a function of the difference between
the dependent variable, whose name is stored in the global macro ML_y1 by ml
and the linear predictor or the mean `xb´. The final command replace `lj´=
ln(`f´) places the log-likelihood functions into the variable `lj´ as required by
the calling program.

Note that all variables defined within a program to be used with ml should be
of storage type double to enable ml to estimate accurate numerical derivatives.
Scalars should be used instead of local macros to hold constants as scalars have
higher precision.

The command

```
ml model lf mixing0 (xb: y=)
```

specifies the method as lf and the program to evaluate the log-likelihood
contributions as mixing0.

The dependent and explanatory variable are specified by the "equation" in
brackets. The name before the colon, xb, is the name of the equation, the
variable name after the colon, y, is the dependent variable, and the variable
names after the '=' give the explanatory variables for the linear predictor. No
explanatory variables are given, so a constant only model will be fitted.

As a result of this model definition, ML_y1 will be equal to y and in mixing0 `xb´ will be equal to the intercept parameter (the mean) that is going to be estimated by ml.

Now maximize the likelihood using

```
ml maximize
```

giving the results shown in Display 13.1. The program converged in three itera-

```
initial:       log likelihood = -1397.9454
alternative:   log likelihood = -1160.3298
rescale:       log likelihood = -197.02113
Iteration 0:   log likelihood = -197.02113
Iteration 1:   log likelihood = -146.79076
Iteration 2:   log likelihood = -146.78981
Iteration 3:   log likelihood = -146.78981

                                           Number of obs   =        100
                                           Wald chi2(0)    =          .
Log likelihood = -146.78981                Prob > chi2     =          .

------------------------------------------------------------------------------
        y |     Coef.   Std. Err.      z    P>|z|     [95% Conf. Interval]
----------+-------------------------------------------------------------------
    _cons |   5.002311         .1   50.023   0.000     4.806314    5.198307
------------------------------------------------------------------------------
```

Display 13.1

tions and the maximum likelihood estimate of the mean is equal to the sample mean of 5.002311. If we are interested in observing the change in the estimate of the mean in each iteration, we can use the trace option in the ml maximize command. The Wald chi2(0) output is empty because we have not specified a likelihood with which to compare the current model. We can suppress this uninformative output using ml maximize with the noheader option.

We now need to extend the program step by step until it can be used to estimate a mixture of two Gaussians. The first step is to allow the standard deviation to be estimated. Since this parameter does not contribute linearly to the linear predictor used to estimate the mean, we need to define another linear predictor by specifying another equation with no dependent variable in the ml model command as follows:

```
ml model lf mixing1 (xb: y=) (lsd:)
```

The new equation has the name lsd, has no dependent variable and the linear predictor is simply a constant. A short-form of the above command is

```
ml model lf mixing1 (xb: y=) /lsd
```

We intend to use lsd as the log-standard deviation. Estimating the log of the standard deviation will ensure that the standard deviation itself is positive. We now need to modify the function mixing0 so that it has an additional argument for the log standard deviation.

```
capture program drop mixing1
program define mixing1
    version 6.0
    args lj xb ls

    tempvar f s

    quietly gen double `s´ = exp(`ls´)
    quietly gen double `f´ = /*
    */ exp(-.5*(($ML_y1-`xb´)/`s´)^2)/(sqrt(2*_pi)*`s´)
    quietly replace `lj´ = ln(`f´)
end
```

In mixing1, s is now the name of a variable instead of a scalar because, in principal, the linear predictor `ls´, defined in the ml model command, may contain covariates (see exercises) and hence differ between observations. Making these changes and using

ml maximize

gives the output shown in Display 13.2.

```
initial:        log likelihood = -1397.9454
alternative:    log likelihood = -534.94948
rescale:        log likelihood = -301.15405
rescale eq:     log likelihood = -180.56802
Iteration 0:    log likelihood = -180.56802
Iteration 1:    log likelihood = -146.65085
Iteration 2:    log likelihood = -146.56478
Iteration 3:    log likelihood = -146.56469
Iteration 4:    log likelihood = -146.56469
-------------------------------------------------------------------------
       y |     Coef.    Std. Err.      z     P>|z|     [95% Conf. Interval]
---------+---------------------------------------------------------------
xb       |
   _cons |   5.002311   .1047816   47.740   0.000     4.796942    5.207679
---------+---------------------------------------------------------------
lsd      |
   _cons |   .0467083   .0707107    0.661   0.509     -.091882    .1852987
-------------------------------------------------------------------------
```

Display 13.2

The standard deviation estimate is obtained by taking the exponential of the estimated coefficient of _cons in equation lsd. Instead of typing display exp(0.0467083), we can use following the syntax for accessing coefficients and their standard errors:

```
display [lsd]_b[_cons]
```

$$\boxed{.04670833}$$

```
display [lsd]_se[_cons]
```

$$\boxed{.07071068}$$

We can also omit "_b" from the first expression and compute the required standard deviation using

```
display exp([lsd][_cons])
```

$$\boxed{1.0478163}$$

This is smaller than the sample standard deviation from **summarize** because the maximum likelihood estimate of the standard deviation is given by

$$\hat{\sigma} = \sqrt{\frac{1}{n}\sum_{i=1}^{n}(y_i - \overline{y})^2} \tag{13.3}$$

where n is the sample size, whereas the factor $\frac{1}{n-1}$ is used in **summarize**. Since n is 100 in this case, the maximum likelihood estimate must be blown up by a factor of $\sqrt{100/99}$ to obtain the sample standard deviation:

```
display exp([lsd][_cons])*sqrt(100/99)
```

$$\boxed{1.053095}$$

The program can now be extended to estimate a mixture of two Gaussians. In order to be able to test the program on data from a known distribution, we will generate a sample from a mixture of two Gaussians with variances of 1 and means of 0 and 5 and with mixing probabilities $p_1 = p_2 = 0.5$. This can be done in two stages; first randomly allocate observations to groups (variable z) with probabilities p_1 and p_2 and then sample from the different component densities according to group membership.

```
clear
set obs 300
set seed 12345678
gen z = cond(uniform()<0.5,1,2)
gen y = invnorm(uniform())
replace y = y + 5 if z==2
```

We now need five equations, one for each parameter to be estimated: μ_1, μ_2, σ_1, σ_2 and p_1 (since $p_2 = 1 - p_1$). As before, we can ensure that the estimated standard deviations are positive by taking the exponential inside the program. The mixing proportion p_1 must lie in the range $0 \leq p_1 \leq 1$. One way of ensuring this is to interpret the linear predictor as representing the log odds (see Chapter 6) so that p_1 is obtained from the linear predictor of the log odds, lo1 using the transformation $1/(1+\exp(-\text{lo1}))$. The program mixing1 now becomes

```
capture program drop mixing2
program define mixing2
    version 6.0
    args lj xb xb2 lo1 ls1 ls2

    tempvar f1 f2 p s1 s2

    quietly gen double `s1´ = exp(`ls1´)
    quietly gen double `s2´ = exp(`ls2´)
    quietly gen double `p´ = 1/(1+exp(-`lo1´))

    quietly gen double `f1´ = /*
    */ exp(-.5*((\$ML_y1-`xb´)/`s1´)^2)/(sqrt(2*_pi)*`s1´)
    quietly gen double `f2´ = /*
    */ exp(-.5*((\$ML_y1-`xb2´)/`s2´)^2)/(sqrt(2*_pi)*`s2´)
    quietly replace `lj´ = ln(`p´*`f1´ + (1-`p´)*`f2´)
end
```

Stata simply uses the value 0 as the initial value for all parameters. When fitting a mixture distribution, it is not advisable to start with the same initial value for both means. Therefore, the initial values should be set. This can be done using the ml init command between the ml model and ml maximize commands as follows:

```
ml model lf mixing2 (xb1: y=) /xb2 /lo1 /lsd1 /lsd2
ml init 1 6 0 0.2 -0.2, copy
ml maximize, noheader
```

The first two values will be the initial values for the means, the third the initial log odds, and the third and fourth the initial values of the logs of the standard deviations. The results are shown in Display 13.3

The standard deviations are therefore estimated as

```
display exp([lsd1][_cons])
```

```
.97290582
```

and

```
initial:        log likelihood = -746.68918
rescale:        log likelihood = -746.68918
rescale eq:     log likelihood = -676.61764
Iteration 0:    log likelihood = -676.61764  (not concave)
Iteration 1:    log likelihood = -630.71045
Iteration 2:    log likelihood = -625.88875
Iteration 3:    log likelihood = -622.23405
Iteration 4:    log likelihood = -622.21162
Iteration 5:    log likelihood = -622.21162
```

y	Coef.	Std. Err.	z	P>|z|	[95% Conf.	Interval]
xb1						
_cons	-.0457452	.0806427	-0.567	0.571	-.203802	.1123117
xb2						
_cons	4.983717	.0863492	57.716	0.000	4.814476	5.152958
lo1						
_cons	.1122415	.1172032	0.958	0.338	-.1174725	.3419555
lsd1						
_cons	-.027468	.0627809	-0.438	0.662	-.1505164	.0955804
lsd2						
_cons	-.0173444	.0663055	-0.262	0.794	-.1473008	.1126121

Display 13.3

```
display exp([lsd2][_cons])
```

.9828052

and the probability of membership in group 1 is

```
display 1/(1 + exp(-[lo1][_cons]))
```

.52803094

Alternative estimates of mixing proportion and means and standard deviations, knowing group membership in the sample, are obtained using

```
table z, contents(freq mean y sd y)
```

z	Freq.	mean(y)	sd(y)
1	160	-.0206726	1.003254
2	140	5.012208	.954237

The estimated parameters agree quite closely with the true parameters and

with those estimated by knowing the group membership. The true proportion
was 0.5 and the realized proportion is $160/300 = 0.53$, also very close to the
maximum likelihood estimate.

The standard errors of the estimated means are given in the regression table in Display 13.3 (0.081 and 0.086). We can estimate the standard errors
of the standard deviations and of the probability from the standard errors
of the log standard deviations and log odds using the delta-method (see for
example Dunn, 1989). According to the delta-method, if $y = f(x)$, then ap-
proximately, $se(y) = |f'(x)|se(x)$ where $f'(x)$ is the first derivative of $f(x)$ with
respect to x evaluated at the estimated value of x. For the standard deviation,
$sd = \exp(lsd)$, so that, by the delta method,

$$se(sd) = sd * se(lsd). \tag{13.4}$$

For the probability, $p = 1/(1 + \exp(-lo))$, so that

$$se(p) = p(1 - p) * se(lo). \tag{13.5}$$

The approximate standard errors for the standard deviations are therefore

```
scalar x = exp([lsd1][_cons])
display [lsd1]_se[_cons]*scalar(x)
```

$$\boxed{.06107994}$$

```
scalar x = exp([lsd2][_cons])
display [lsd2]_se[_cons]*scalar(x)
```

$$\boxed{.06516543}$$

and for the probability

```
scalar x = 1/(1 + exp(-[lo1][_cons]))
display [lo1]_se[_cons]*scalar(x*(1-x))
```

$$\boxed{.0292087}$$

In the above commands, the function **scalar()** was used to make sure that
x is interpreted as the scalar we have defined and not as a variable we may
happen to have in the dataset. This was not necessary in the program **mixing2**
because there we used a temporary name; temporary names look something
like **__0009J4** and are defined deliberately to be different from any variable
name in the dataset.

We can now apply the same program to the age of onset data. The data can
be read in using

```
infile y using onset.dat, clear
gen ly=ln(y)
label variable y "age of onset of schizophrenia"
```

A useful graphical display of the data is a histogram found from

```
graph y, hist bin(10) xlab ylab gap(3)
```

which is shown in Figure 13.1.

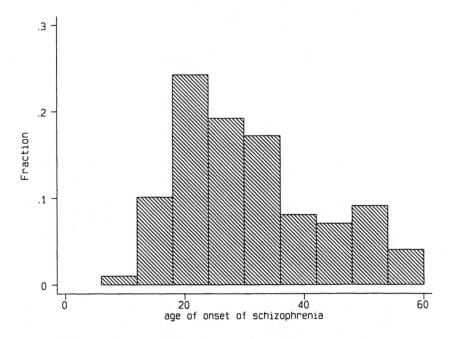

Figure 13.1 *Histogram of age of onset of schizophrenia in women.*

It seems reasonable to use initial values of 20 and 45 for the two means. In addition, we will use a mixing proportion of 0.5 and log standard deviations of 2 as initial values.

```
ml model lf mixing2 (y: y=) /xb2 /lo1 /lsd1 /lsd2
ml init 20 45 0 2 2, copy
ml maximize, noheader
```

The output is given in Display 13.4. The means are estimated as 24.8 and 46.4, the standard deviations are estimated as 6.5 and 7.1 for groups 1 and 2, respectively, and the mixing proportions are estimated as 0.74 and 0.26. The approximate standard errors may be obtained as before; see exercises. The histogram strongly suggests that there are two subpopulations. In order to test this more formally, we could also fit a single normal distribution and compare the likelihoods. Wolfe (1971) suggests, on the basis of a limited simulation study, that the difference in minus twice the log-likelihood for a model with c

```
initial:        log likelihood = -391.61146
rescale:        log likelihood = -391.61146
rescale eq:     log likelihood = -391.61146
Iteration 0:    log likelihood = -391.61146  (not concave)
Iteration 1:    log likelihood = -374.66666  (not concave)
Iteration 2:    log likelihood = -374.36519
Iteration 3:    log likelihood = -373.93993
Iteration 4:    log likelihood = -373.67089
Iteration 5:    log likelihood = -373.66896
Iteration 6:    log likelihood = -373.66896
```

y	Coef.	Std. Err.	z	P>|z|	[95% Conf. Interval]	
xb1						
_cons	24.79772	1.133545	21.876	0.000	22.57601	27.01942
xb2						
_cons	46.44685	2.740866	16.946	0.000	41.07485	51.81885
lo1						
_cons	1.034415	.3697502	2.798	0.005	.3097178	1.759112
lsd1						
_cons	1.877698	.1261092	14.889	0.000	1.630529	2.124868
lsd2						
_cons	1.955009	.25692	7.609	0.000	1.451455	2.458563

Display 13.4

components compared with a model with $c + 1$ components has approximately a χ^2 distribution with $2\nu - 2$ degrees of freedom where ν is the number of extra parameters in the $c + 1$ component mixture. The log-likelihood of the current model may be accessed using $e(11)$. We store this in a local macro

```
local 11 = e(11)
```

and fit the single normal model using the program **mixing1** as follows:

```
ml model lf mixing1 (xb: y=) /lsd
ml init 30 1.9, copy
ml maximize, noheader
```

with the result shown in Display 13.5

We can now compare the log-likelihoods using the method proposed by Wolfe:

```
local chi2 = 2*(`11'-e(11))
display chiprob(4,`chi2')
```

```
.00063994
```

This test confirms that there appear to be two subpopulations.

```
initial:        log likelihood = -429.14924
rescale:        log likelihood = -429.14924
rescale eq:     log likelihood = -429.14924
Iteration 0:    log likelihood = -429.14924
Iteration 1:    log likelihood = -383.83957
Iteration 2:    log likelihood =  -383.3985
Iteration 3:    log likelihood = -383.39585
Iteration 4:    log likelihood = -383.39585
```

y	Coef.	Std. Err.	z	P>|z|	[95% Conf.	Interval]
xb						
_cons	30.47475	1.169045	26.068	0.000	28.18346	32.76603
lsd						
_cons	2.453747	.0710669	34.527	0.000	2.314458	2.593035

Display 13.5

13.4 Exercises

1. Create a do-file with the commands necessary to fit the mixture model.

2. Add commands to the end of the do-file to calculate the standard deviations and mixing probability and the standard errors of these parameters. What are the standard errors of the estimates for the age of onset data?

3. Simulate values from two normals trying out different values for the various parameters. Do the estimated values tend to lie within two estimated standard errors from the 'true' values?

4. Use program mixing1 to fit a linear regression model to the slimming data from Chapter 5 using only status as the explanatory variable. Compare the standard deviation with the root mean square error obtained using the command regress.

5. Use the same program again to fit a linear regression model where the variance is allowed to differ between the groups defined by status. Is there any evidence for heteroscedasticity? How do the results compare with those of sdtest?

6. Extend the program mixing2 to fit a mixture of 3 normals and test this on simulated data (hint: use transformations $p1 = 1/d$, $p2 = \exp(lo1)/d$, and $p3 = \exp(lo2)/d$ where $d = 1 + \exp(lo1) + \exp(lo2)$) and test it on simulated data.

Appendix: Answers to Selected Exercises

Chapter 1

Assuming that the data are stored in the directory c:\user,

2. ```
cd c:\user
insheet using test.dat, clear
```

4. ```
label define s 1 male 2 female
label values sex s
```

5. ```
gen id=_n
```

6. ```
rename v1 time1
rename v2 time2
rename v3 time3
```

   or

   ```
for var v1-v3 \ num 1/3: rename X timeY
```

7. ```
reshape long time, i(id) j(occ)
```

8. ```
egen d=mean(time), by(id)
replace d=(time-d)^2
```

9. ```
drop if occ==3&id==2
```

## Chapter 2

1. ```
table depress, contents(mean weight)
```

2. ```
for var iq age weight: table life, contents(mean X sd X)
```

3. ```
search mann
help signrank
```

4. ```
ranksum weight, by(life)
```

5. ```
gen iq1=iq if life==2
gen iq2=iq if life==1
label variable iq1 "no"
label variable iq2 "yes"
graph iq1 iq2 age, s(dp) xlabel ylabel jitter(2) l1("IQ")
```

6. Save the commands in the "Review" window and edit the file using the Do-file Editor or any texteditor, e.g., Notepad. Add the commands given in the do-file template in Section 1.9 and save the file with the extension *.do*. Run the file by typing the command do `filename`.

Chapter 4

1. ```
infile bp11 bp12 bp13 bp01 bp02 bp03 using bp.raw
```

2. ```
sort drug
graph bp, box by(drug)
sort diet
graph bp, box by(diet)
sort biofeed
graph bp, box by(biofeed)
```

4. ```
sort id
save bp
infile id age using age.dat, clear
sort id
merge id using bp
anova bp drug diet biofeed age, continuous(age)
```

## Chapter 5

1. ```
anova resp cond*status status cond, sequential
```

2. ```
gen dcond=cond-1
gen dstat=status-1
gen dinter=dcond*dstat
regress resp dcond dstat dinter
```

3. ```
xi: regress resp i.cond*i.status
```

4. ```
char cond[omit] 2
char status[omit] 2
xi: regress resp i.cond*i.status
```

## Chapter 6

1. ```
ologit outc therapy sex [fweight=fr], table
```

2.(a) ```
ologit depress life
```

  (b) ```
replace life=life-1
logistic life depress
```

3. Even if we use very lenient inclusion and exclusion criteria,

```
sw logistic life depress anxiety iq sex sleep, /*
     */ pr(0.2) pe(0.1) forward
```

only depress is selected. If we exclude depress from the list of candidate variables, anxiety and sleep are selected.

4.
```
bprobit pres tot ck
predict predp
graph predp prop ck, c(s.)
```

Chapter 7

1.
```
xi: glm resp i.cond i.status, fam(gauss) link(id)
local dev1=e(deviance)
xi: glm resp i.cond, fam(gauss) link(id)
local dev2=e(deviance)
local ddev=`dev2´-`dev1´
/* F-test equivalent to anova cond status, sequential */
local f=(`ddev´/1)/(`dev1´/31)
disp `f´
disp fprob(1,31,`f´)
/* difference in deviance */
disp `ddev´
disp chiprob(1, `ddev´)
```

2.
```
reg resp status, robust
ttest resp, by(status) unequal
```

3.
```
gen cleth=class*ethnic
glm days class ethnic cleth, fam(poiss) link(log)
```

4.
```
glm days class ethnic if stres<4, fam(poiss) link(log)
```
or, assuming the data has not been sorted,
```
glm days class ethnic if _n~=72, fam(poiss) link(log)
```

5.
```
gen abs=cond(days>=14,1,0)
glm abs class ethnic, fam(binomial) link(logit)
glm abs class ethnic, fam(binomial) link(probit)
```

6.
```
logit abs class ethnic, robust
probit abs class ethnic, robust
bs "logit abs class ethnic" "_b[class] _b[ethnic]", reps(500)
bs "probit abs class ethnic" "_b[class] _b[ethnic]", reps(500)
```

Chapter 8

1.
```
graph dep1-dep6, box by(group)
```

2. We can obtain the mean over visits for subjects with complete data using
   the simple command (data in 'wide' form)

   ```
   gen av2 = (dep1+dep2+dep3+dep4+dep5+dep6)/6
   ```

   The t-tests are obtained using

   ```
   ttest av2, by(group)
   ttest av2, by(group) unequal
   ```

3. ```
   egen avg = rmean(dep1-dep6)
   egen sd = rsd(dep1-dep6)
   gen stdav= avg/sd
   ttest stdav, by(group)
   ```

4.(a) ```
   gen diff = avg-pre
       ttest diff, by(group)
   ```

  (b) ```
   anova avg group pre, continuous(pre)
   ```

## Chapter 9

2.(a) ```
   infile subj group pre dep1 dep2 dep3 dep4 dep5 dep6 /*
       */ using depress.dat
   mvdecode _all, mv(-9)
   reshape long dep, i(subj) j(visit)
   reg dep group pre visit, robust cluster(subj)
   ```

  (b) ```
   bs "reg dep group pre visit" "_b[group] _b[pre] _b[visit]", /*
       */ cluster(subj) reps(500)
   ```

5. ```
   infile subj y1 y2 y3 y4 treat baseline age using chemo.dat
   reshape long y, i(subj) j(week)
   expand 2 if week==1
   sort subj week
   qui by subj: replace week=0 if _n==1
   replace y=baseline/4 if week==0
   gen post=week!=0
   xi: xtgee y i.treat*i.post age, i(subj) t(week) corr(exc) /*
           */ family(pois) scale(x2) eform
   ```

Chapter 10

1. ```
   infile v1-v2 using estrogen.dat, clear
   gen str8 _varname="ncases1" in 1
   replace _varname="ncases0" in 2
   xpose,clear
   gen conestr=2-_n
   reshape long ncases, i(conestr) j(casestr)
   ```
   ```

```
   expand ncases
   sort casestr conestr
   gen caseid=_n
   expand 2
   sort caseid
   quietly by caseid: gen control=_n-1 /*control=1,case=0 */
   gen estr=0
   replace estr=1 if control==0&casestr==1
   replace estr=1 if control==1&conestr>0
   gen cancer=control==0
   clogit cancer estr, group(caseid) or
```

2. ```
   table exposed, contents(sum num sum py)
   iri 17 28 2768.9 1857.5
   ```

3. ```
   xi: poisson num i.age*exposed, e(py) irr
   testparm IaX*
   ```

   The interaction is not statistically significant at the 5% level.

4. ```
   infile subj y1 y2 y3 y4 treat baseline age using chemo.dat
   reshape long y, i(subj) j(week)
   expand 2 if week==1
   sort subj week
   qui by subj: replace week=0 if _n==1
   replace y=baseline if week==0
   gen post=week!=1
   gen ltime=log(cond(week==0,8,2))
   xi: xtgee y i.treat*i.post age , i(subj) t(week) corr(exc) /*
           */ family(pois) scale(x2) offset(ltime)
   ```

Chapter 11

1. We consider anyone still at risk after 450 days as being censored at 450 days and therefore need to make the appropriate changes to status and time before running Cox regression.

   ```
   replace status=0 if time>450
   replace time=450 if time>450
   stset time, failure(status)
   stcox dose prison clinic
   ```

2. The model is fitted using

   ```
   gen dose_cat=0 if dose~=.
   replace dose_cat=1 if dose>=60
   replace dose_cat=2 if dose>=80
   xi: stcox i.dose_cat i.prison i.clinic, bases(s)
   ```
   ```

The survival curves for no prison record, clinic 1 are obtained using

```
gen s0 = s if dose_cat==0
gen s1 = s^(exp(_b[Idose__1]))
gen s2 = s^(exp(_b[Idose__2]))
graph s0 s1 s2 time, sort c(JJJ) s(...) xlab ylab gap(3)
```

Note that the baseline survival curve is the survival curve for someone whose covariates are all zero. If we had used `clinic` instead of `i.clinic` above, this would have been meaningless; we would have had to exponentiate `s0`, `s1` and `s2` by `_b[clinic]` to calculate the survival curves for clinic 1.

3. Treating dose as continuous:

```
gen clindose=clinic*dose
stcox dose prison clinic clindose
```

Treating dose as categorical:

```
xi: stcox i.dose_cat*i.clinic i.prison
testparm IdX*
```

5.
```
gen tpris=prison*(t-504)/365.25
stcox dose prison tpris, strata(clinic)
lrtest, saving(0)
stcox dose prison, strata(clinic)
lrtest
```

Chapter 12

1.
```
factor 1500-r4000, pc cov
```

3.
```
capture drop pc*
factor 1500-r4000, pc
score pc1-pc5
graph pc1-pc5, matrix
```

5.
```
infile str10 town so2 temp manuf pop wind precip days using usair.dat
factor temp manuf pop wind precip days,pc
score pc1 pc2
graph pc2 pc1, symbol([town]) ps(150) gap(3) xlab ylab
```

6.
```
regress so2 pc1 pc2
gen line=pc1*_b[pc2]/_b[pc1]
graph pc2 line pc1, s([town]i) c(.1) ps(150) gap(3) xlab ylab
```

Chapter 13

2.
```
scalar x = exp([lsd1][_cons])
disp "st. dev. 1 = " scalar(x) ", standard error = " /*
```

```
      */ [lsd1]_se[_cons]*scalar(x)

scalar x = exp([lsd2][_cons])
disp "st. dev. 2 = " scalar(x) ", standard error = " /*
      */ [lsd2]_se[_cons]*scalar(x)

scalar x = 1/(1 + exp(-[lo1][_cons]))
disp "probability = " scalar(x) ", standard error = "/*
      */ [lo1]_se[_cons]*scalar(x*(1-x))
```

giving the results

```
st. dev. 1 = 6.5384366, standard error = .82455744
st. dev. 2 = 7.0639877, standard error = 1.8148844
probability = .7377708, standard error = .07153385
```

4. The only thing that is different from fitting a normal distribution with constant mean is that the mean is now a linear function of **status** so that the **ml model** command changes as shown below

```
infile cond status resp using slim.dat, clear
ml model lf mixing1 (xb: resp = status) /lsd
ml maximize, noheader
```

In linear regression, the mean square error is equal to the sum of squares divided by the degrees of freedom, n-2. The maximum likelihood estimate is equal to the sum of squares divided by n. We can therefore get the root mean square error for linear regression using

```
disp exp([lsd][_cons])*sqrt(_N/(_N-2))
```

Note that the standard error of the regression coefficients need to be corrected by the same factor, i.e.,

```
disp [xb]_se[status]*sqrt(_N/(_N-2))
```

Compare this with the result of

```
regress resp status
```

5. Repeat the procedure above but replace the **ml model** command by

```
ml model lf mixing1 (resp = status) (lsd: status)
```

The effect of **status** on the standard deviation is significant (p=0.003) which is not too different from the result of

```
sdtest resp, by(status)
```

6. ```
capture program drop mixing3
program define mixing3
```
```

```
version 6.0
args lj xb xb2 xb3 lo1 lo2 ls1 ls2 ls3

tempvar f1 f2 f3 p1 p2 p3 s1 s2 s3 d

quietly gen double `s1´ = exp(`ls1´)
quietly gen double `s2´ = exp(`ls2´)
quietly gen double `s3´ = exp(`ls3´)
quietly gen double `d´ = 1 + exp(`lo1´) + exp(`lo2´)
quietly gen double `p1´ = 1/`d´
quietly gen double `p2´ = exp(`lo1´)/`d´
quietly gen double `p3´ = exp(`lo2´)/`d´


quietly gen double `f1´ = /*
*/ exp(-.5*((${ML_y1}-`xb´)/`s1´)^2)/(sqrt(2*_pi)*`s1´)
quietly gen double `f2´ = /*
*/ exp(-.5*((${ML_y1}-`xb2´)/`s2´)^2)/(sqrt(2*_pi)*`s2´)
quietly gen double `f3´ = /*
*/ exp(-.5*((${ML_y1}-`xb3´)/`s3´)^2)/(sqrt(2*_pi)*`s3´)
quietly replace `lj´ = ln(`p1´*`f1´ + `p2´*`f2´ + `p3´*`f3´)
end


clear
set obs 300
set seed 12345678
gen z = uniform()
gen y = invnorm(uniform())
replace y = y + 5 if z<1/3
replace y = y + 10 if z<2/3&z>=1/3
ml model lf mixing3 (xb1: y=) /xb2 /xb3 /lo1 /lo2 /lsd1 /lsd2 /lsd3
ml init 0 5 10 0 0 0 0 0, copy
ml maximize, noheader trace
```

References

Aitkin, M. 1978. The analysis of unbalanced cross-classifications (with discussion). *Journal of the Royal Statistical Society, A*, **41**, 195–223.

Binder, D. A. 1983. On the variances of asymptotically normal estimators from complex surveys. *International Statistical Review*, **51**, 279–292.

Boniface, D. R. 1995. *Experimental Design and Statistical Methods*. London: Chapman & Hall.

Caplehorn, J., & Bell, J. 1991. Methadone dosage and the retention of patients in maintenance treatment. *The Medical Journal of Australia*, **154**, 195–199.

Chatterjee, S., & Price, B. 1991. *Regression Analysis by Example (2nd Edition)*. New York: Wiley.

Clayton, D., & Hills, M. 1993. *Statistical Models in Epidemiology*. Oxford: Oxford University Press.

Collett, D. 1991. *Modelling Binary Data*. London: Chapman & Hall.

Collett, D. 1994. *Modelling Survival Data in Medical Research*. London: Chapman & Hall.

Cook, R. D. 1977. Detection of influential observations in linear regression. *Technometrics*, **19**, 15–18.

Cook, R. D. 1979. Influential observations in linear regression. *Journal of the American Statistical Association*, **74**, 164–174.

Cook, R. D., & Weisberg, S. 1982. *Residuals and Influence in Regression*. London: Chapman & Hall.

Diggle, P. J., Liang, K-Y, & Zeger, S. L. 1994. *Analysis of Longitudinal Data*. Oxford: Oxford University Press.

Dunn, G. 1989. *Design and Analysis of Reliability Studies*. Oxford: Oxford University Press.

Efron, B., & Tibshirani, R. 1993. *An Introduction to the Bootstrap*. London: Chapman & Hall.

Everitt, B. S. 1993. *Cluster Analysis*. London: Edward Arnold.

Everitt, B. S. 1994. *Statistical Methods for Medical Investigations*. London: Edward Arnold.

Everitt, B. S. 1995. The analysis of repeated measures: A practical review with examples. *The Statistician*, **44**, 113–135.

Everitt, B. S. 1996. An introduction to finite mixture distributions. *Statistical Methods in Medical Research*, **5**, 107–127.

Everitt, B. S., & Der, G. 1996. *A Handbook fo Statistical Analyses using SAS*. London: Chapman & Hall.

Everitt, B. S., & Dunn, G. 1991. *Applied Multivariate Analysis*. London: Edward Arnold.

Everitt, B. S., & Rabe-Hesketh, S. 1997. *The Analysis of Proximity Data*. London: Edward Arnold.

Goldstein, H. 1995. *Multilevel Statistical Models*. London: Edward Arnold.

Gould, W. 1999. *Estimating a Cox model with continuously time-varying parameter*. Stata FAQ.

Gould, W., & Sribney, W. 1999. *Maximum Likelihood Estimation with Stata*. College Station, TX: Stata Press.

Gregoire, A. J. P., Kumar, R., Everitt, B. S., Henderson, A. F., & Studd, J. W. W. 1996. Transdermal oestrogen for the treatment of severe post-natal depression. *The Lancet*, **347**, 930–934.

Hamilton, L. C. 1998. *Statistics with Stata 5*. Belmont, CA: Duxbury Press.

Hand, D. J., & Crowder, M. 1996. *Practical Longitudinal Data Analysis*. London: Chapman & Hall.

Hand, D. J., Daly, F., Lunn, A. D., McConway, K. J., & Ostrowski, E. 1994. *A Handbook of Small Data Sets*. London: Chapman & Hall.

Holtbrugge, W., & Schumacher, M. 1991. A comparison of regression models for the analysis of ordered categorical data. *Applied Statistics*, **40**, 249–259.

Hotelling, H. 1933. Analysis of complex statistical variables into principal components. *Journal of Education Psychology*, **24**, 417–441.

Jackson, J. E. 1991. *A User's Guide to Principal Components*. New York: Wiley.

Jennrich, R. I., & Schluchter, M. D. 1986. Unbalanced repeated measures models with unstructured covariance matrices. *Biometrics*, **42**, 805–820.

Lachenbruch, P., & Mickey, R. M. 1986. Estimation of error rates in discriminant analysis. *Technometrics*, **10**, 1–11.

Lewine, R. R. J. 1981. Sex differences in schizophrenia: timing or subtypes? *Psychological Bulletin*, **90**, 432–444.

Liang, K-Y., & Zeger, S. L. 1986. Longitudinal data analysis using generalised linear models. *Biometrika*, **73**, 13–22.

Mallows, C. L. 1986. Some comments on C_p. *Technometrics*, **15**, 661–675.

Manly, B. F. J. 1997. *Randomisation, Bootstrap and Monte Carlo Methods in Biology*. London: Chapman & Hall.

Matthews, J. N. S., Altman, D. G., Campbell, M. J., & Royston, P. 1990. Analysis of seriel measurements in medical research. *British Medical Journal*, **300**, 230–235.

Maxwell, S. E., & Delaney, H. D. 1990. *Designing Experiments and Analysing Data*. California: Wadsworth.

McCullagh, P., & Nelder, J. A. 1989. *Generalized Linear Models*. London: Chapman & Hall.

McKay, R. J., & Campbell, N. A. 1982a. Variable selection techniques in discriminant analysis. I Description. *British Journal of Mathematical and Statistical Psychology*, **35**, 1–29.

McKay, R. J., & Campbell, N. A. 1982b. Variable selection techniques in discriminant analysis. II Allocation. *British Journal of Mathematical and Statistical Psychology*, **35**, 30–41.

McLachlan, G. J., & Basford, K. E. 1988. *Mixture Models; Inference and Application to Clustering*. New York: Marcel Dekker, Inc.

Nelder, J. A. 1977. A reformulation of linear models. *Journal of the Royal Statistical Society, A*, **140**, 48–63.

Pearson, K. 1901. On lines and planes of closest fit to points in space. *Philosophical Magazine*, **2**, 559–572.

Rawlings, J. O. 1988. *Applied Regression Analysis*. California: Wadsworth Books.

Rothman, K. J. 1986. *Modern Epidemiology*. Boston: Little, Brown & Company.

Sackett, D. L., Haynes, R. B., Guyatt, G. H., & Tugwell, P. 1991. *Clinical Epidemiology*. Massachusetts: Little Brown & Company.

Sokal, R. R., & Rohlf, F. J. 1981. *Biometry*. San Francisco: W. H. Freeman.

Sprent, P. 1993. *Applied Nonparametric Statistical Methods*. Chapman & Hall: P. Sprent.

Sribney, B. 1998. *What are the advantages of using robust variance estimators over standard maximum-likelihood estimators in logistic regression*. Stata FAQ.

Stata Getting Started. 1999. *Getting Started with Stata for Windows*. College Station, TX: Stata Press.

Stata Graphics Manual. 1999. *Stata Graphics Manual*. College Station, TX: Stata Press.

Stata Manuals 1-4. 1999. *Stata Reference Manuals*. College Station, TX: Stata Press.

Stata User's Guide. 1999. *Stata User's Guide*. College Station, TX: Stata Press.

Thall, P. F., & Vail, S. C. 1990. Some covariance models for longitudinal count data with overdispersion. *Biometrics*, **46**, 657–671.

Wedderburn, R. W. M. 1974. Quasilikelihoodfunctions, generalised linear models and the Gauss-Newton method. *Biometrika*, **61**, 439–47.

Winer, B. J. 1971. *Statistical Principles in Experimental Design*. New York: McGraw-Hill.

Wolfe, J. H. 1971. *A Monte Carlo study of the sampling distribution of the likelihood ratio for mixtures of multinormal distributions*. Technical Bulletin, vol. STB72-2. San Diego: Naval Personnel and Training Research Laboratory.

Index

temporary variable, 182
time-varying covariate, 162, 166
trace option, 183
trim() option, 41
Type I sums of squares, *see* sequential
  sums of squares
Type III sums of squares, *see* unique
  sums of squares

unbalanced, 70
unique sums of squares, 72
updating Stata, 22–23
using(0) option, 82

variance components, 125
variance inflation factors, 45

weights, 8, 15, 100

xlabel option, 16

ylabel option, 16
yline(0) option, 30
yscale() option, 115